Table of Contents

Introduction

This text gives intermediate-level adult English learners the language skills needed to access, evaluate, and use technology in everyday life. The units address technology applied in research, consumer-related, and workplace situations. Research applications include automated library systems and basic internet search functions. Consumer applications include card readers, automated teller machines, and self-serve postal equipment. Technology applied in workplace situations includes communication systems, cash registers, data entry operations, and shipping and tracking systems. Each unit features a different setting, and involves a realistic situation in which a non-native speaker might encounter the technology for the first time.

Unit Opener

Each unit begins with an illustration and a few warm-up questions to introduce the setting and the type of technology to be covered. Then there is a story or dialogue that describes a situation in which a person must use a new kind of everyday technology.

Look and Listen

This section presents new vocabulary in a visual context. After listening to the new language and looking at the visual presentation, students may be asked to identify, match, or sequence vocabulary items, or to act out the steps in a procedure. The *English for Technology* cassette and the tapescript in the Teacher's Guide provide a copy of the language used to present the new vocabulary.

Focus on Listening, Focus on Communication

In these segments, students will hear conversations related to using the new technology. These conversations include problem-solving language, expressions needed to get more information or clarify a procedure, ways to make suggestions or offer help, and other useful sentences and phrases for dealing with technology. The final listening task will involve targeted listening, in which students are asked to listen and fill in specific words or phrases from the tape. Speaking activities that follow will provide opportunities for oral practice. Further practice of the communication skills includes games, readings, role plays and other activities found later in the unit.

Real-Life Reading, Focus on Reading

The reading sections of each unit are essential for developing the skills needed for accessing technology. *Real-Life Reading* indicates a reading lesson that involves reading natural material as it appears on a piece of equipment, a computer screen, or an authentic information source, such as a chart or a poster. *Focus on Reading* lesson segments provide additional work on reading skills, using a more traditional source, such as a story or a paragraph from a brochure.

Focus on Information, Coping Skills, Community Assignment

These unique sections of each lesson present the information and language skills required for working with a particular form of technology, and then challenge students to use all the skills learned in a task outside the classroom. They are designed as preparation for real-world use of the reading, communication, and technology access skills presented in the unit. It is our hope that students will find *English for Technology* to be a tool for success in today's world of ever-increasing technological advancement into everyday life.

English for Technology: Scope and Sequence, Units 1-5

Unit	Technology Goal	Technology Access Skills	Communication Skills	Grammatical Structures
1	*Pay at the Pump* Use pay-point pumps	Read digital displays Follow written instructions Follow oral instructions	Ask for help Ask about steps in a procedure Use short questions to ask for clarification	Imperatives Two-word verbs
2	*Can I Use Plastic?* Use supermarket check-out machines	Use a numerical keypad Clarify oral instructions Read informational brochures Fill out a supermarket card application	Ask for information Ask about procedures Repeat to ask for clarification	Modals Question formation with modals and *do*
3	*I Need Cash Qucik!* Use ATM machines	Follow written instructions Select from on-screen choices to carry out a procedure Interpret banking information	Ask hypothetical questions Talk about consequences Listen for new vocabulary in context	Questions with *What if* Past participles
4	*Was that a Hamburger Combo?* Help customers at a drive-through window	Follow oral instructions Take fast-food orders Listen to garbled speech Make change	Acknowledge oral instructions Repeat to confirm information Ask for repetition	Negative imperatives Compound sentences with *when, if, as, after*
5	*Twenty on eight, Please* Use a computerized cash register	Use a complex keyboard Interpret abbreviations Follow oral instructions	Respond to customer orders Indicate understanding of oral instructions Ask about meaning Ask about function	Review

English for Technology: Scope and Sequence, Units 6-10

Unit	Technology Goal	Technology Access Skills	Communication Skills	Grammatical Structures
6	*Self Service at the Post Office* Use a post office scale and vending machine	Interpret information screens Page forward and back Select from on-screen choices to find information Read labels and codes	Make suggestions Offer assistance Suggest possibilities Talk about potential outcomes	Modals *could* and *would*
7	*Springfield Public Library* Use an online catalog at a library	Follow on-screen instructions Use search commands Scan a list of entries Interpret library information Interpret new vocabulary in context	Ask for advice Talk about alternatives	Simple Past for unspecified time in the past Present Perfect for unspecified time in the past
8	*How May I Direct Your Call?* Use office machines and process orders	Answer multi-line phones in an office Transfer calls Use voice mail Enter purchase order data on a computer Use keyboard commands Demonstrate a procedure	Describe equipment problems Use polite language Give instructions Ask a favor Write notes to remember a procedure	Verb+object+infinitive *She asked me to ...*
9	*Packing and Shipping* Use a warehouse tracking system for packing and shipping	Interpret transfer orders Enter information on a shipping computer Use mouse commands Interpret computer terminology Select from on-screen menus	Talk about steps carried out in a procedure Help identify errors in a procedure Train another worker on a procedure	Impersonal *you* Simple Past for identification of completed actions
10	*I Found It on the Internet* Use the internet	Read complex displays Interpret internet terminology Identify steps in an internet search Identify on-screen menus Identify and select appropriate on-screen choices	Follow written instructions	Review

Pay at the Pump

Where do you buy gas for your car? Do you use self-service or full-service?
How do you pay for your gas? Do you ever use a credit card?

Got Gas?

Sam is in the car with his girlfriend, Wendy. They are going to see a movie, and they are in a
hurry. Wendy is driving. She pulls into a gas station, pulls a credit card out of her wallet, and
gets out of the car. She puts her card somewhere on the pump, presses some buttons, and fills
the gas tank in a few minutes. Sam is watching with interest. He is a new immigrant from a
small town, and he likes to learn about the use of modern technology in everyday life.

Answer the following questions about the story.

1. Who is Wendy?
2. Where are Sam and Wendy going?
3. Who is driving?
4. Where does Wendy keep her credit cards?
5. What does Wendy do to get gas?
6. How long does it take to get gas?
7. What is Sam doing?
8. How does Sam feel about the use of technology in everyday life?

Look and Listen 1 👁 👁 👂

A. Look at the illustrations and listen to the tape.

1.
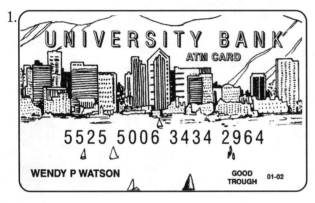

2.

3.
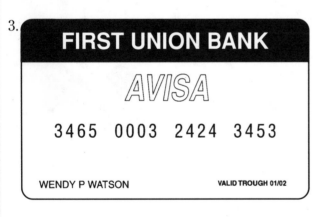

4.

B. Listen to the tape. Number the sentences to match the illustrations.

_____ This is my Uniworld card. It's a credit card, too.

__1__ This is my ATM card from the bank. It's a debit card.

_____ This one is my driver's license. It's also an ID card.

_____ This one is my AVISA card. It's a credit card.

Focus on Communication 1

So Many Cards!

When Wendy left the car to pump her gas, her wallet dropped on the car seat and a lot of cards fell out. Sam picked them up for her. After she got back into the car, he asked some questions about them. Wendy answered his questions.

Words	Meanings
plastic	plactic cards, such as credit cards or debit cards
convenient	easy to use
borrow	get money from someone or a company and pay it back later
interest	money that you pay back in addition to the money you borrow
purchase	buy something
purchases	things that you buy
pay the bill	pay for goods and/or services that you receive
cash	paper money or coins (currency)

A. Listen to Sam and Wendy's conversation and check true (T) or false (F).

	T	F
1. Wendy doesn't need all the plastic.	_____	_____
2. An AVISA card is a debit card.	_____	_____
3. An ATM card is a debit card.	_____	_____
4. When Wendy uses her credit card, she is borrowing money.	_____	_____
5. When Wendy uses her ATM card, she is borrowing money.	_____	_____

B. Answer the following questions.

1. What kind of card is an ATM card?
2. When does Wendy use an ATM card?
3. What is the difference between a debit card and a credit card?
4. How does Sam like to pay for his purchases?
5. How can Wendy avoid paying interest on her credit card purchases?

C. Listen to the tape carefully. Fill in the blanks with the words you hear. Listen to the full conversation as many times as necessary.

Sam: _____ this one?

Wendy: _____ an ATM card.

Sam: What _____ you use it _____?

Wendy: I use it _____ get cash from an ATM machine.

Sam: _____ this other one?

Wendy: Oh, this is my AVISA. _____ a credit card.

Sam: _____ the difference between these two?

Wendy: _____ the ATM the money comes from my bank account, and _____
                the credit card I pay the bill later.

Sam: You _____ every time you use a credit card you are borrowing money?

Wendy: That's _____.

Sam: _____ that mean that you have to pay interest on what you buy with
                your credit card?

Wendy: Yes, it does.

D. Practice the dialogue above with another student. Then switch roles.

E. Look at the pictures on Page 2 and practice asking another student about the cards. Use the questions below. Then switch roles.

Example: What's this card?

                What do you use it for?

                What's the difference between _____ and _____?

                You mean _____?

                Does that mean _____?

Look and Listen 2

Sam is very excited today. He now has a credit card for Uniworld Gas and wants to use it for the first time to pay for his gas at the pump. He stops at a Uniworld gas station. He parks his car at a self-service island and prepares to get gas.

A. Look at the pictures and listen to the tape.

1.

2.

3.
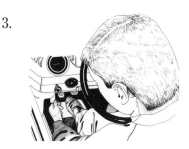

4.

5.

6.

B. Listen to the tape and number the sentences below in the correct order.

_____ Remove the gas cap.

_____ Get out of your car.

__1__ Turn off the engine.

_____ Lock the door.

_____ Take out your credit card.

_____ Pull up the lever to release the gas tank cover.

C. Act out the sentences as another student gives the instructions. Then switch roles.

Look and Listen 3 👁 👁 👂

A. Look at the pictures and listen to the tape.

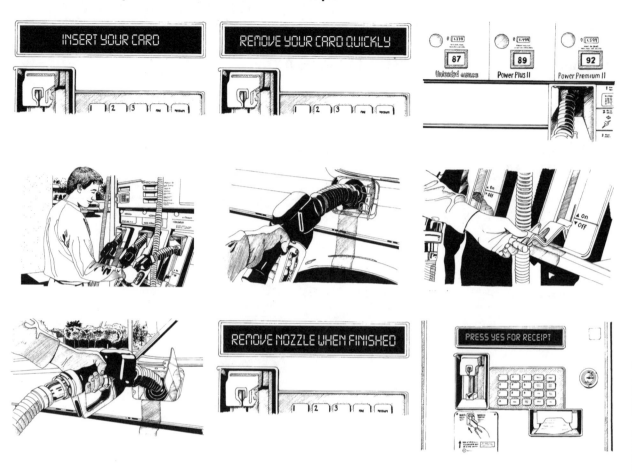

B. Listen for the order of the steps. Then use the lines on the next page to copy the instructions below in the correct order.

press YES for a receipt

insert your card

pick up the nozzle

lift up the lever on the pump

select the gas octane and press the button

remove your card quickly

when you're finished, remove the nozzle

put the nozzle into the tank

squeeze the handle to begin fueling

Buying Gas with Your Card

First _____ .

Now _____ .

Afterward _____ .

After that, _____ .

Next _____ .

Then _____ .

After that, _____ .

And then _____ .

Finally, _____ .

C. Act out the sentences above as another student says the steps. Then switch roles.

Focus on Communication 2

Can You Help Me with This?

Sam tried to pay at the pump and get gas, but he had a problem. He called the gas station attendant, and the attendant came to help him. He is talking to the attendant now.

A. Listen for the general ideas and then check true (T) or false (F). Listen as many times as necessary.

		T	F
1.	Sam wants to pay with his credit card.	_____	_____
2.	He needs to leave the card in the machine.	_____	_____
3.	He needs to choose the octane before he can pump gas.	_____	_____
4.	Sam wasn't able to get gas the first time he tried.	_____	_____

B. Answer the following questions. Work with the teacher or with a partner.

1. Why does Sam call the attendant?

2. What is the first step in paying with a credit card?

3. What is the next step?

4. What does Sam do to choose the octane?

5. Why wasn't Sam able to get gas the first time he tried?

6. Is the attendant helpful? What does he tell Sam to do?

C. Listen to Sam's questions and fill in the blanks.

Sam: Excuse me, can you tell me _____ _____ pay with my credit card?

Sam: You _____ I should put it in here?

Sam: Which way _____ the card face?

Sam: This _____?

Sam: Okay. _____ next?

Sam: How _____ I do that?

Sam: Like _____?

Sam: _____ happened? There's no gas.

Sam: This _____?

D. Listen to the attendant's answers and fill in the blanks. Listen as many times as necessary.

1. Sam: You mean I should put it in here?

   Attendant: _____ right.

2. Sam: Which way should the card face?

   Attendant: _____ way. Just like the picture here.

   Sam: This way?

   Attendant: No, the _____ way.

   Sam: Oh, I see.

3. Sam: Okay. What's next?

   Attendant: _____ you need to choose the right octane.

   Sam: How do I do that?

   Attendant: _____ the button for the one you want.

   Sam: Like this?

   Attendant: _____.

4. Sam: What happened? There's no gas.

   Attendant: _____ you have to lift up the lever on the pump.

   Sam: This one?

   Attendant: _____.

E. Talk to another student. First ask how to do the next step in pumping your gas, and then ask for clarification. Switch roles and answer the questions.

Example: A: Can you tell me how to pay with my credit card?

   B: Sure. You need to insert your card in the pump.

   A: This way?

   B: Yes, that's right.

STUDENT A
Can you tell me how to . . .

- pay with my credit card?
- insert my card?
- release the gas tank cover?
- get the gas to begin fueling?
- select the octane?

STUDENT B
Sure. You have to . . .

Yes. You need to . . .

STUDENT A
You mean . . .?

   This way?

   Like this?

   This one?

STUDENT B
Yes.

Right.

That's right.

No, the other way.

No, like this.

No, the other one.

Vocabulary

A. Match the words with the pictures. Write the correct letter on the line.

a.

b.

c.

d.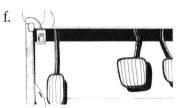

e.

f.

_____ 1. nozzle

_____ 2. gas tank cover

_____ 3. lever for gas tank cover release

_____ 4. gas pump lever

_____ 5. gas gauge

_____ 6. lever for gas release on the nozzle handle

B. Identify the correct words to fill in the blanks.

gas cap	release	remove	insert	nozzle	lever	gas gauge

1. The _____ is really low. We need to get gas soon.

2. He pulled the _____ to release the gas tank cover.

3. He put the _____ into the tank.

4. He could not remove the _____ before releasing the cover.

5. First _____ the gas tank cover and then _____ the gas cap.

6. To pay with a credit card, you need to _____ your card first.

Focus on Grammar

A. Write the letter of the meaning on the right in the blank next to the two-word verb on the left that it matches.

Example: *Put in your card* has the same meaning as *Insert your card*, so they are a match.

Two-word Verbs

_b_ put in

_____ get in

_____ take out

_____ turn off

_____ let go

_____ take off

_____ pull up

Meanings

a. remove

b. insert

c. enter

d. release

e. remove

f. lift

g. stop

B. Fill in the blanks with matching two-word verbs from the list below.

put in	take out	pull up	turn off	let go (of)	get in

1. Be sure to _____ the engine when you pull in to a gas station.
   stop

2. When you _____ the lever, the gas tank cover will open.
   lift

3. If you want to pay at the pump, the first step is to _____ your credit card.
   insert

4. The instructions on the gas pump say to _____ your card quickly.
   remove

5. When you are finished pumping your gas, you need to _____ the handle on the
   release

   nozzle before you return the nozzle to the pump.

6. Don't forget to put the gas cap on before you _____ the car.
   enter

C. Guess which two-word verbs are opposites and draw a line to match them.

turn off   put in

take out   press down

get out   get in

lift up   put on

take off   put back

pick up   turn on

D. Write the letter of the instruction on the right in the blank next to the instruction on the left that has the same meaning.

Example: *Lift up the lever on the gas pump* can mean the same as *Lift it up*, so they are a match.

_____ Take off the gas cap.   a. Lift it up.

__a__ Lift up the lever on the gas pump.   b. Put it in here.

_____ Put your card in here.   c. Pick it up.

_____ Turn off the engine.   d. Turn it off.

_____ Press down the lever.   e. Press it down.

_____ Pull up the gas tank cover release lever. f. Take it off.

_____ Pick up the nozzle.   g. Pull it up.

Role Play

Imagine that you are asking for help at the gas pump. One person asks for help. The other person explains. The first person asks for clarification. The second person gives more information and explains all the steps.

Example: Student A: Can you tell me how to pay with my credit card?
             Student B: First insert your card.
             Student A: How do I do that?
             Student B: (Continue the conversation.)

Concentration Game

A. How to make sentences

Match the words on the left with the words on the right. See how many sentences you can make by combining the two groups.
Example: *Take out your credit card.*

turn off	engine
take out	credit card
put in	ATM card
pick up	nozzle
get in	car
lift up	lever
press	button
squeeze	handle

B. How to prepare the game

- Organize students in groups of two to four.
- Divide an $8\frac{1}{2}$-by-11-inch sheet of paper into sixteen sections and cut them apart.
- Write one of the above words or two-word verbs on each section of paper to make cards.
- Arrange the cards in four rows with four cards in each row.
- Place the cards face down on the table.

C. How to play the game

- Players take turns.
- The first player turns over any two cards.
- If the two cards go together in a sentence, say the sentence correctly and keep the two cards.
- If they don't go together, put them back on the table face down.
- Each player turns over two cards and tries to make a match.
- When there are no cards left on the table, all players count their cards.
- The player with the most cards is the winner.

Focus on Reading

Safety First

Words	Meanings
run errands	go out to do different things that are necessary, such as shopping, picking up dry cleaning, and so forth
drive-thru	(or drive-through or drive-up) a place where you can receive service without leaving your car

Wendy has a sister. Her name is Sara. Sara has two children: a three-year-old son and a six-month-old daughter. Whenever she runs errands, she has to take her children in the car with her. That's why she likes drive-thru restaurants and banks with drive-up windows. She always goes to gas stations where she can pay at the pump with her credit card or ATM card. Of course, if she goes to the full-service island, she doesn't have to get out of the car, but full-service costs a lot more.

Sara is very careful about the safety of her children. She pays at the pump because she doesn't want to leave the children in the car while she walks to the cashier's booth, pays, and walks back. Also, it is usually difficult to get the children out of the car and take them with her. Before she gets out of the car to pump gas, Sara rolls up the windows. Then she locks the car. "You can never be too careful," she says.

A. Read the sentences and check true (T) or false (F).

		T	F
1.	Sara is Wendy's sister.	___	___
2.	Sara has two sons.	___	___
3.	She prefers to use pay-point gas pumps.	___	___
4.	She usually gets gas at the full-service island.	___	___

B. Circle the letter of the best answer.

1. Sara prefers drive-thru restaurants and banks because
   a. They provide faster service.
   b. They cost less.
   c. They are safer to use.
   d. Both a and c.

2. She doesn't use regular self-service gas pumps because
   a. She is too tired to walk and talk to the cashier.
   b. She doesn't want to leave the kids in the car alone.
   c. She doesn't have enough time.
   d. She doesn't know how to use them.

3. When Sara says, "You can never be too careful," she probably means
   a. It is good to be careful all the time.
   b. Even if you are very careful, it's never enough.
   c. It is good to be careful, but not all the time.
   d. Both a and b.

C. Discuss these questions with a partner or group.

1. What is Sara's concern? How does she deal with the problem? Is Sara too careful? Why or why not?
2. Name five errands that you run every week. Tell when and how you do them.
3. Name two fast-food restaurants and two banks with drive-thru services. How is the food or the service at those places?
4. What is the difference between a self-service and full-service island in a gas station? Which costs less? Which provides more service?
5. Which is faster? Paying at the pump or at the cashier's booth? Have you ever tried paying at the pump?
6. Why is it safer for Sara to pay at the pump?
7. Why does Sara lock her car when she gets out to pay for gas?
8. What are the advantages and disadvantages of using credit cards? Is it safe to use credit cards? Why or why not?

Coping Skills

When Sam went to the gas station and tried to pay at the pump, what went wrong?
Did Sam solve his problem?

Put a ✓ next to the steps he took to solve his problem. Give examples.

( )   read the instructions on the pump

( )   followed written instructions

( )   asked for help

( )   asked about steps in the procedure

( )   asked questions to clarify the instructions

( )   followed spoken instructions

With a group, discuss the coping skills above.

What is the most important skill you learned in this unit? Which ones will you use in the future? In what situations will you use them?

Community Assignment

Get gas at a computerized pump and bring the receipt to class. If you don't drive a car, go with a friend who drives and help your friend buy gas at a computerized pump. Then bring the receipt to class. Show the receipt to the class and explain where you bought the gas, how you did it, what questions you asked, how much gas you bought, and how much you paid per gallon.

Can I Use Plastic?

Where is Mitra? What is she doing? Where do you usually shop for food?
How do you usually pay? Do you ever use *plastic* to pay for your groceries?

At the Supermarket

Mitra is standing in line to pay for her groceries at the market. She has been waiting in line a
really long time today. It seems like everyone in her line needs to write a check, ask the
checker a lot of questions, look for an ID, or look for discount coupons before paying. It's
taking such a long time! Usually Mitra pays cash, and sometimes she even goes to the Express
Lane if she has only a few items. Today she doesn't have enough cash to pay for all her
groceries, and she needs to write a check, so she has to wait in the regular line.

Answer these questions about the story.

1. Why is Mitra standing in line?
2. Does she usually have to wait a long time?
3. Which line is Mitra standing in today? Why?
4. Why is it taking such a long time?
5. What are discount coupons?

Look and Listen 1 👁 👁 👂

A. Look at Mitra's check below and answer the questions.

Mitra Amini	90-1765	101
1455 S. Maple St.	1234	
Anytown, CA 91728		

DATE _Jan. 22, 1999_

PAY TO THE
ORDER OF _Happy Market_ $ _43.67_

Forty-three and 67/100 DOLLARS

Gold Rush Bank
WEST CITY BRANCH
1023 EAST BLVD.
EAST CITY, CALIFORNIA 90036

MEMO _____ _Mitra Amini_

⑆122000473⑆ 0349627833⑈ 0101

1. What is the date today?
2. What is the name of the market?
3. How much is she paying for her groceries?
4. What is Mitra's last name?
5. Where does she sign her name?

B. Listen to the instructions for writing a check. Write the number of each step in the correct place on the blank check below.

Mitra Amini
1455 S. Maple St.
Anytown, CA 91728

90-1765
1234
102

DATE _____

PAY TO THE
ORDER OF _____ $_____

_____ DOLLARS

Gold Rush Bank
WEST CITY BRANCH
1023 EAST BLVD.
EAST CITY, CALIFORNIA 90036

MEMO _____

⑆122000473⑆ 0349627833⑈ 0102

C. Listen to the instructions for writing a check again. Fill in the missing words below.

1. _____ the date.

2. Write the _____ of the market.

3. Write in _____ the amount you are _____.

4. _____ out the amount you are paying.

5. Draw a _____ up to the word *DOLLARS*.

6. Now _____ the check.

Focus on Listening 1

Writing Checks

A. Listen to the tape and circle the letter for each item you hear.

1. a. Happy Market   b. Davis and Associates   c. Water and Power Co.

2. a. Super Drugs   b. Pacific Gas and Oil Co.   c. American Management Co.

3. a. $30.48   b. $13.48   c. $14.48

4. a. $12.22   b. $10.72   c. $12.72

5. a. $40.98   b. $40.88   c. $14.98

6. a. $745.00   b. $645.00   c. $635.00

7. a. Nineteen and 14/100   b. Ninety and 14/100

   c. Nineteen and 40/100

8. a. Thirty-one and 99/100   b. Thirty-one and 95/100

   c. Forty-one and 95/100

9. a. Twenty-seven and 15/100   b. Twenty-seven and 50/100

   c. Twenty-seven and 60/100

10. a. Eight hundred eighty-six and 33/100   b. Eight hundred eighty-six and 43/100

    c. Eight hundred ninety-six and 33/100

B. Listen to Mitra say how she is making out her checks. Write the information she says on the blank checks below.

1.

Mitra Amini	90-1765 / 1234 103
1455 S. Maple St.	
Anytown, CA 91728	DATE _____
PAY TO THE	
ORDER OF _____ $_____
_____ DOLLARS
Gold Rush Bank	
WEST CITY BRANCH	
1023 EAST BLVD.	
EAST CITY, CALIFORNIA 90036	
MEMO _____ _____
⑆122000473⑆ 0349627833⑊ 0103	

2.

Mitra Amini	90-1765 / 1234 104
1455 S. Maple St.	
Anytown, CA 91728	DATE _____
PAY TO THE	
ORDER OF _____ $_____
_____ DOLLARS
Gold Rush Bank	
WEST CITY BRANCH	
1023 EAST BLVD.	
EAST CITY, CALIFORNIA 90036	
MEMO _____ _____
⑆122000473⑆ 0349627833⑊ 0104	

3.

Mitra Amini	90-1765 / 1234 105
1455 S. Maple St.	
Anytown, CA 91728	DATE _____
PAY TO THE	
ORDER OF _____ $_____
_____ DOLLARS
Gold Rush Bank	
WEST CITY BRANCH	
1023 EAST BLVD.	
EAST CITY, CALIFORNIA 90036	
MEMO _____ _____
⑆122000473⑆ 0349627833⑊ 0105	

C. Talk with another person and compare your checks. Then discuss your answers with your teacher.

1. Did you write the correct information?
2. After spelling out the amount of money, did you draw a line all the way to the word *DOLLARS*?
3. What did Mitra mean when she said, "Now I have to write these down in my check register"? Why is it important to do that?
4. If you make a mistake when you are writing a check, do you know what to do?
5. What do you think you can use the MEMO line for? Can you think of some examples for Mitra's checks?

Look and Listen 2

The next time Mitra went to the market, she saw that many people were using plastic—credit cards, debit cards, and so forth—to pay for their groceries. She thought that would be much faster, so she asked if she could use her automated teller machine (ATM) card. The checker said yes. Then he showed her a special card reader where she could insert her ATM card and her personal identification number, or PIN.

A. Listen to the instructions and look at the card reader pictured below. Draw a circle around the part of the card reader that is needed for each action.

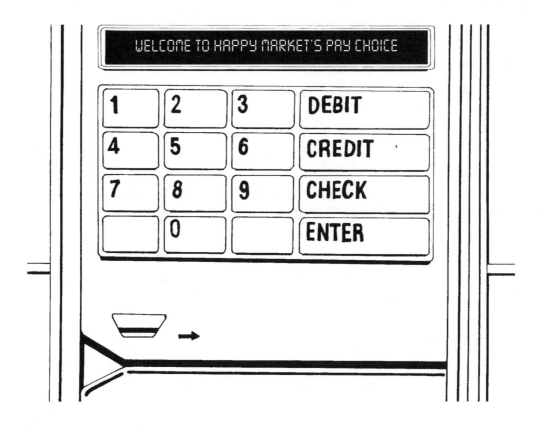

B. Read the steps for using the card reader and number them in the correct order.

SELECT PAYMENT TYPE. _____

THANK YOU. _____

SWIPE YOUR CARD. _____

ENTER AMOUNT AND PRESS ENTER. _____

ENTER YOUR PIN AND PRESS ENTER. _____

Focus on Communication 1

Can I Use Plastic?

A. Listen to the conversation between Mitra and the checker at the checkstand. Then check true (T) or false (F).

	T	F
1. Mitra wants to pay with a credit card.	_____	_____
2. She needs to give her card to the checker.	_____	_____
3. At first, when she swipes her card, the machine doesn't respond.	_____	_____
4. She has to select her type of payment first.	_____	_____
5. She has to press ENTER before she enters her PIN.	_____	_____
6. Next she has to enter the amount of her groceries.	_____	_____
7. She has to press ENTER after she enters the amount.	_____	_____
8. Mitra thinks paying with an ATM card is efficient.	_____	_____

B. Listen to the way Mitra asks the checker to explain, and fill in the blanks. Then practice the conversation with a partner.

Checker: Now enter your PIN.
Mitra: _____ _____?
Checker: That's your secret number.
Mitra: Oh, yeah. (beep, beep, beep, beep)
Checker: Now enter the amount.
Mitra: _____ _____?
Checker: Yes. Your total is $12.98. Enter it and then press ENTER.
Mitra: Oh, Okay. (beep, beep, beep, beep... beep)

Four Convenient Ways to Pay

Words	Meanings
major	big, important
advantage	something helpful; an extra benefit
double advantage	two advantages
discount	lower price
checking account	bank account with checks for making payments
apply	ask for, fill out an application for something

When she left the market, Mitra picked up this brochure. It explains all the ways of paying at Happy Market. She took it home to read.

ATM cards are welcome at all Happy Market branches. Use your ATM card, and payment will come automatically from your bank account. It's fast, easy, and convenient.

Happy Market now accepts all major credit cards. Shop now and pay later. Simply press the button and use the card of your choice.

With the Happy Shopper's Club card, you'll get all the advantages that come with club membership. You'll receive special members-only discounts on thousands of items. Apply for your card today and save!

With Happy Market electronic checking, you will enjoy a double advantage. You will receive the discount prices of the Happy Shopper's Club and automatic payment from your checking account. You'll never have to write a check at Happy Market—just use your card!

Read the brochure and select one of the payment choices to fill in the blanks.

1. If you want payment for your purchases to come from your bank account automatically, use a _____ card.

2. If you want to enjoy discount prices, you should get a _____ _____ _____ card.

3. If you want a double advantage of low prices and payment that comes directly from your checking account, you should apply for _____ _____ with your Happy Shopper's Club card.

4. If you want to shop now and pay later, use your _____ card.

The Happy Shopper's Club Card

Mitra went back to Happy Market today. She picked up three bottles of juice, two packages of meat, and a few other things. When she went to the checkstand, the checker asked her if she had a Happy Shopper's Club card.

A. Listen to the conversation and check true (T) or false (F).

		T	F
1.	Mitra is a Happy Shopper's Club member.	_____	_____
2.	Club members receive discounts on some items.	_____	_____
3.	Mitra wants to be a member.	_____	_____
4.	She can use electronic checking today.	_____	_____
5.	She needs to fill out an application.	_____	_____
6.	The checker will help her fill out the application.	_____	_____

B. Listen to the conversation between Mitra and the checker again and fill in the blanks.

Checker: _____ _____ have a Happy Shopper's Club card?

Mitra: No, I don't think so. _____ that?

Checker: It's a Happy Market discount card. If you use the card, _____ _____ get an automatic discount on our specials.

Mitra: On specials?

Checker: Yes. For example, this juice is $1.79 on special, and the meat is $3.99 a pound.

Mitra: How _____ _____ get a Happy Shopper's Club card?

Checker: You _____ _____ fill out an application and become a club member.

Mitra: _____ _____ _____ _____ pay with a Happy Shopper's Club card?

Checker: No, you get the lower prices when you swipe your Happy Shopper's Club card, but you can pay any way you like.

Mitra: So _____ _____ _____ _____ swipe my ATM card and then my Happy Shopper's Club card?

Checker: No. If you want to, _____ _____ apply for electronic checking, too. Then your Happy Shopper's Club card works like an ATM card. You can use it to pay here at Happy Market, and _____ _____ _____ to write checks.

Mitra: _____ _____ apply and get the lower prices today?

Checker: _____ _____ get the Happy Shopper's Club discount card today, but electronic checking will take longer.

Mitra: _____ _____ give me an application?

Checker: Sure, here you are. _____ _____ fill it out over there, please?

Mitra: Yes. Thank you for your help.

C. Practice the conversation between Mitra and the checker with a partner.

D. Rewrite the following questions with the words in the correct order.

1. an /use / card /I / ATM / Can ?

   _____

2. Happy Shopper's Club / Do/ have / card / you / a ?

   _____

3. Should / again / slide / I / in / card / my ?

   _____

4. price / much / is / How / regular / the ?

   _____

5. How / Happy Shopper's Club / I / get / a / card / do ?

   _____

6. have to / Happy Shopper's Club / Do / pay / I /a / with / card ?

   _____

7. swipe / ATM / and / have to / card / then / my / my / Happy Shopper's Club / Do / card/ I ?

   _____

8. today / apply / get / Can / lower / I / the / and / prices ?

   _____

9. me / application / Could / an / you / give ?

   _____

10. you / Would / out / fill / it / over / please / there ?

    _____

E. Talk to another student. The first person asks questions using the words below. The second person answers. Then switch roles.

Student A: Can I . . .?   **Student B:** Yes.

Should I . . .?   No.

Do I have to . . .?   (Then give more information.)

F. Talk to another student. The first person makes requests with the words below. The second person answers. Then switch roles.

Student A: Could you . . .?   **Student B:** Sure.

Would you . . .?   Okay.

               Of course.

               (Then do what was requested.)

Focus on Grammar

A. Study the following words and their meanings. Then write a sentence or a request with each one on the lines below.

Words	Meanings	Examples
can	able, possible	You can pay with a check, credit card, or ATM card.
should	good idea	You should be very careful to keep your PIN secret.
have to	necessary	You don't have to show your ID when you use an ATM card or a credit card.
will	future activity	Mitra will fill out an application for a Happy Shopper's Club card.
could	polite request	Could you get some milk at the market please?
would	very polite request	Would you please take my groceries to the car?

1. _____

2. _____

3. _____

4. _____

5. _____

6. _____

B. Now share your examples with a partner and with the teacher.

Happy Shopper's Club Membership

Mitra decided to apply for a Happy Shopper's Club card. She wants to use the card instead of writing checks at the supermarket. She can pay quickly by swiping her card. She can also use the Express Lane if she has twelve items or less, and she will get special discounts that are for members only.

If Mitra needs to fill out an application now, she will get a discount card right away. Because she wants electronic checking also, she needs to include a sample check with her application. The check will show the necessary information about her checking account at the bank. Then she will sign the application form, giving Happy Market permission to take money directly from her account when she uses the card to pay for her groceries.

After she sends in her application and receives her card, Mitra needs to go back to the market to choose her personal identification number (PIN). This is very important because only she will know the PIN, and only she will be able to use the card to pay for her groceries. She will even be able to get twenty dollars cash each time she goes to the market. This is going to be an efficient way to pay at the supermarket, and a convenient way to get extra cash from her checking account, too.

A. Discuss the following vocabulary words with another student or with your teacher.

to print	to write with clear, separate letters, not with connected handwriting
to have privileges	to be able to do something special
a notice	an announcement or letter that gives information
to allow	to permit
to authorize	to give permission
financial institution	bank or credit union
valid	good, up-to-date

A preprinted check   *A voided check*

To tape something

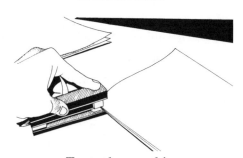

To staple something

B. Read the application and then check true (T) or false (F).

Happy Shopper's Club Application

Please complete and sign this form to join the Happy Shopper's Club and enjoy valuable club member savings. If you wish to have electronic check privileges, please include your preprinted voided check. You will receive a notice to bring to your local store that will allow you to select your personal identification number (PIN) for electronic checking.

Please Print Clearly

Applicant's Last Name   First Name   M.I.

Street Address   Apt. No.   Area Code   Home Phone Number

City   State   Zip Code   Area Code   Work Phone Number

Birth Date _____

Applicant's Driver License   State   Mo.   Day   Year
or State ID

**For electronic check privileges, tape preprinted voided check here.
Do not staple.**

The Happy Shopper's Club Customer Agreement:
By signing this application, and after using my club card at any Happy Market, I agree to the following:

For each use of my Happy Shopper's Club card as an electronic check, I authorize my financial institution to pay from my checking account to Happy Market the amount of the purchase and/or cash received.

Applicant's signature (MUST BE SIGNED TO BE VALID)

X _____   Date _____

                                                            T   **F**

1. If Mitra wants electronic checking, she needs to tape
   a voided check on her application.   _____   _____

2. A personal identification number is called a PIN.   _____   _____

3. Mitra can be a Happy Shopper's Club member, even if she
   doesn't want electronic checking.   _____   _____

4. If she includes a voided check, Mitra gives Happy Market
   permission to take money directly from her bank account.   _____   _____

C. Now read the application for a Happy Shopper's Club card again from the beginning, and fill it out for yourself.

Vocabulary

A. Match the words in bold below with the definitions to the right. Write the letter of the correct definition in the blank.

_____ You can pay by **check**.

a. to look at something carefully

_____ **Check** your receipt to make sure it's correct.

b. a small paper that is filled out and signed to pay for something from a bank account

_____ She is standing in line at the **checkstand**.

c. the place where you pay for your groceries

_____ He is a **checker**.

d. a cashier in a supermarket

B. Fill in the blanks with words from the list below.

convenient	efficient	notice	authorize	advantage	discount	valid

1. You will receive a _____ when your new card is ready.

2. One _____ of being a member is that you can get lower prices.

3. It is a good idea to shop for _____ prices.

4. Writing checks is not very _____ when you're in a hurry.

5. With your signature on the application, you _____ the market to take money directly from your checking account.

6. Using a credit card is _____, but sometimes the interest is expensive.

7. An application for electronic checking is only _____ if a voided check is attached and the application is signed and dated.

Role Play

Imagine that you are a checker in a supermarket. As different people come through the line at your checkstand, help them pay by check, ATM card, or credit card. Other students will act out the parts of customers. You will greet them, ring up their groceries, ask them how they would like to pay, answer their questions, thank them, and say good-bye when they leave.

Coping Skills

When Mitra wanted to pay with her ATM card, did she know what to do? How did she learn more about how to do it?

Put a ✓ next to the steps she took to find out about different ways to pay at the supermarket. Give examples.

( )   asked for help

( )   asked for clarification

( )   asked for additional information

( )   read the instructions on the card reader

( )   followed written instructions

( )   followed spoken instructions

( )   repeated the instructions back for clarification

( )   got more information from a brochure

With a group, discuss the coping skills above.

What is the most important skill you learned in this unit? Which ones will you use in the future? In what situations will you use them?

Community Assignment

Go to several supermarkets, or have members of your group go to different supermarkets. Find out the ways to pay at the market, what kinds of cards you can use there, and whether they have a "Club" card. Pick up an application for club membership, electronic checking, or other services that are available. Bring the information and the applications to share with your group. Then decide which market you think is the most convenient.

I Need Cash—Quick!

What are the men at the table doing? What are the people down the street doing? Have you ever used an ATM machine to get cash?

A Money Machine

Ignacio: Hi, Gonzalo. Sorry I'm late. I had to stop and get some cash on my way.

Gonzalo: Get some cash? You mean you went to the bank?

Ignacio: No, I just used the ATM machine down the street.

Gonzalo: Really? How did you do that?

Ignacio: Oh, it's easy. You just put in your card, enter your PIN, and take the cash.

Gonzalo: And the cash comes out of the machine?

Ignacio: Yes, you just take your cash, your card, and your receipt, and that's it.

Gonzalo: Wow! A money machine!

Answer these questions about the conversation.

1. Why was Ignacio late?
2. Where did he go to get some cash?
3. Was it easy to do?
4. Do you think Gonzalo was impressed?

Look and Listen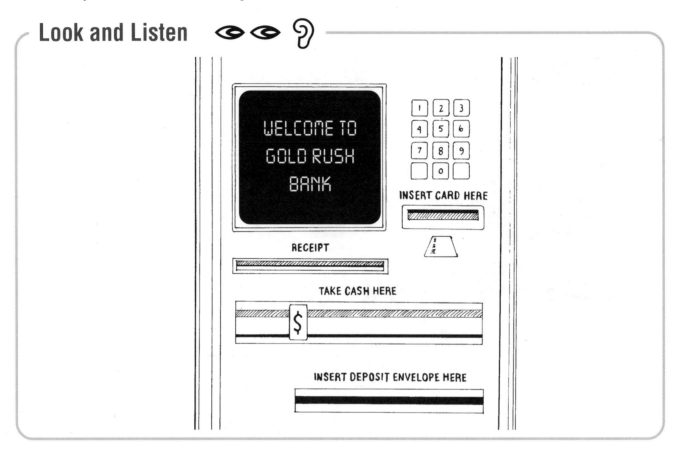

A. Listen to the definition of each word below. Match the letter of the word with the correct definition on the right.

a.	keypad	_____	A piece of paper that shows what you did at the ATM machine.
b.	receipt	_____	The area where you see instructions for using the machine.
c.	deposit	_____	The place where you enter your PIN.
d.	message screen	_____	The place where you insert your card.
e.	cash	_____	Money that you put into your bank account.
f.	card slot	_____	Money that you take out of the ATM machine.

B. Work with another student. The first person asks about where to do something on the ATM machine. The second person answers and uses the picture above to show the place on the machine. Then switch roles.

Example: **Student A:** Where do I insert my card?
  Student B: Here. In the card slot. (pointing to the card slot on the picture)

I'd Like Some Information about the ATM

Gonzalo decided to go to his bank and ask about getting an ATM card so that he could use the automated teller machine. He asked the woman at the front desk who he should talk to, and she directed him to the representative at the New Accounts desk. The representative was very helpful and gave him lots of information. The only problem was that she used so many words that were not familiar to Gonzalo. So what did he do? He just continued the conversation to get the main idea.

A. Listen for the main idea in the conversation between Gonzalo and the bank representative. Then check true (T) or false (F). Listen as many times as necessary.

	T	F
1. Gonzalo can use the ATM only at Gold Rush Bank.	_____	_____
2. He can withdraw money from his checking account or his savings account.	_____	_____
3. He can use the ATM to see how much money he has in the bank.	_____	_____
4. Gonzalo wants to try using the ATM at his bank.	_____	_____

B. Work with a partner. Ask and answer the following questions. Then switch roles.

1. What can Gonzalo do if he doesn't have enough money in his checking account?
2. Is there a fee for using the ATM machine?
3. Is it possible to transfer money from one account to another?
4. If you have an account at a different bank, can you get cash with your Gold Rush ATM card?
5. How do you make a deposit at the ATM?

C. Work with a partner again. Talk about the meaning of the following words. Then check with your teacher to see if you are correct.

select	transaction
checking account	balance
savings account	withdraw
fee	transfer

Making a Transaction at the ATM

Gonzalo is using the ATM machine for the first time. He wants to withdraw sixty dollars in cash from his checking account. He doesn't mind paying the fee for a cash withdrawal. He definitely wants a receipt for his transaction.

A. Read the ATM messages on the screens below and decide what Gonzalo should do after each instruction.

1.

2.

3.

4.

5.

6.

A | HOW MUCH WOULD YOU LIKE TO WITHDRAW? | E
B | $20 $40 | F
C | $60 $80 | G
D | $100   OTHER AMOUNT | H

7.

A | | E
B | YOUR TRANSACTION | F
C | IS BEING PROCESSED | G
D | PLEASE WAIT | H

8.

A | GOLD RUSH BANK WILL DEDUCT A $1.50 FEE | E
B | FOR THE TRANSACTION YOU HAVE CHOSEN | F
C | PRESS "ENTER" IF YOU ACCEPT THIS FEE | G
D | TO EXIT PRESS "CANCEL" | H

9.

A | | E
B | PLEASE TAKE CASH | F
C | DO YOU WANT ANOTHER TRANSACTION? | G
D | ← YES   NO → | H

10.

A | | E
B | PLEASE REMOVE YOUR CARD AND | F
C | TAKE YOUR RECEIPT | G
D | THANK YOU FOR BANKING AT   GOLD RUSH BANK | H

B. Answer the following questions about the ATM messages.

Screen 1:   What will Gonzalo do first?

Screen 2:   Which button will he press?

Screen 3:   What does Gonzalo need to do now? Where will he do it?

Screen 4:   Which button will Gonzalo press?

Screen 5:   Which transaction should he select?

Screen 6:   Which button will he press?

Screen 7:   What should Gonzalo do now?

Screen 8:   What does *deduct* mean? Which button will Gonzalo press now?

Screen 9:   Which one will Gonzalo choose? What else can he do at this time?

Screen 10:   What two things does Gonzalo have to do before he leaves the machine?

C. Read about the situations below. Then look at the ATM screens on the previous pages and write the number of the message screen or screens that fits each situation. After you choose the screen, write down which button Gonzalo should press in each situation. The first one is done for you.

Situations	Screen	Button
1. He does not need a receipt.	2	G
2. He made a mistake when he entered his PIN.		
3. He wants to withdraw $120 from his checking account.		
4. He wants to know how much money he has in his savings account.		
5. He wants to move some money from his savings account into his checking account.		
6. He wants to withdraw money from his savings account.		
7. He wants to make a deposit into his checking account.		
8. He does not want to pay a fee for his transaction.		
9. He wants to make one more transaction at the ATM.		
10. He wants to make a deposit into his savings account.		

Focus on Listening 2

What If My Card Gets Stolen?

The next time Gonzalo saw Ignacio, he asked him about ATM cards again. He told him that he had gotten an ATM card, but he felt a little worried about using it. He was thinking about what could happen if he did not use it carefully. He wondered what would happen if he had any problems with his ATM card or his PIN.

A. Listen for the general ideas and check true (T) or false (F).

		T	F
1.	Gonzalo can use very few Solar System ATM machines.	_____	_____
2.	If your ATM card is stolen, the bank will give you a new card.	_____	_____
3.	If someone else knows your PIN, they can use your card.	_____	_____
4.	It's a good idea to write down your PIN so that you don't forget it.	_____	_____

B. Answer the following questions.

1. What will the bank do if Gonzalo forgets his PIN?

2. Why isn't the personal identification number written on ATM cards?

3. Why is Gonzalo nervous about using the ATM?

C. Listen to the tape carefully. Fill in the blanks with the words you hear.

1.

Gonzalo: I'm a little nervous about using my ATM card. What if _____ _____ a mistake when I'm using the machine?

Ignacio: _____ you'll see a message that asks you to try again. For example, if you key in the wrong PIN, it will ask you to enter your number again.

2.

Gonzalo: What if my _____ _____ stolen?

Ignacio: _____ you should call the bank right away.

Gonzalo: Oh, I see.

3.

Gonzalo: What if _____ _____ my PIN?

Ignacio: _____ you can call the bank and they will send you a new PIN. Sometimes they send you a new card, too.

Gonzalo: That makes sense. I'm glad they're so careful.

Focus on Communication

A. Work with a partner. The first person asks one of the questions below, the second person answers, and the first person responds to that answer. Then switch roles.

STUDENT A

What if . . .

- my card gets stolen?
- I forget my PIN?
- I want to make a deposit?
- I don't have enough money in my checking account?

STUDENT B

Then . . .

- you should . . .
- you can . . .
- you just . . .
- the bank will . . .

STUDENT A

Oh, I see.

That makes sense.

Okay.

I see.

B. Continue asking and answering new *What if?* questions with your partner. Then write down three questions to ask the class.

1. _____

2. _____

3. _____

C. Check your questions with your partner or your teacher. Then ask others in your class the questions and write their answers below.

1. _____

2. _____

Focus on Reading

The ATM Brochure

Words	Meanings
funds	money
access	get service; connect with a computerized service
provide	give; make available
multiple	number that may be divided by another number equally (Example: eighty is a multiple of twenty)
terminate	stop; end
account	report of banking transaction during a period of time

Last week Gonzalo picked up this brochure at the Gold Rush Bank. It explains the types of services that bank customers can access at the ATM instead of going to the bank during regular business hours.

ATM Services

Gold Rush Automated Teller Machine (ATM) Service is available twenty-four hours a day, seven days a week at most locations.

You can use your ATM card to make any of the following transactions:

- Withdraw cash. You may withdraw up to $200 a day in multiples of $20.
- Transfer funds between your linked checking and savings accounts.
- Make balance inquiries.

There is no fee for transactions at Gold Rush Bank's automated teller machines. However, when you access your account from an ATM outside the Gold Rush system, you will have to pay a fee. Each account that you access will be charged a transaction fee by Gold Rush Bank for each cash withdrawal and each balance inquiry that you make at an ATM. This fee will be deducted from the linked account. Our ATMs normally provide a receipt when you complete a transaction. If a receipt cannot be provided, you may continue or terminate your transaction. In addition, ATM transactions are listed on your monthly checking account statement.

A. Read the sentences and check true (T) or false (F).

	T	F
1. Gold Rush ATMs are closed on Sundays.	____	____
2. You can withdraw a maximum of $400 at the ATM.	____	____
3. There is a transaction fee for cash withdrawals at other ATMs.	____	____
4. The transaction fee is deducted from the cash that you receive at the ATM.	____	____
5. ATM transactions will be on your regular bank statement.	____	____

Oops!

Words	Meanings
invalid	not valid; not correct
denied	not accepted
incorrect	not correct
insufficient	not enough

The next time Gonzalo tried to use the ATM, he had trouble completing his transaction. He made a few mistakes when he was trying to withdraw cash from his checking account. But each time he made a mistake, he saw a message on the screen that told him what was wrong and gave him other choices. By following the instructions on the screen he was able to correct his mistakes and complete the transaction.

A. **Read the mistakes that Gonzalo made and write the number for each mistake in front of the message screen that tells him what to do.**

1. He couldn't remember his PIN and took too long to enter it.
2. He entered the wrong PIN.
3. He tried to withdraw eighty dollars from his checking account, but he didn't have enough money in that account.
4. He chose Quick Cash to withdraw cash from his savings account.
5. He decided to withdraw only fifty dollars from his savings account.

Vocabulary

A. Fill in the blanks with words from the list below.

terminate	transfer funds	access	withdraw	fee	multiple	statement	transaction

1. If you don't remember your PIN, you can't _____ your account.

2. If you don't want to continue a transaction, you can _____ it at any time by pressing the CANCEL key.

3. If you don't have enough money in your checking account, you may _____ _____ from savings to checking.

4. You can _____ up to $200 per day at the ATM.

5. When you make a withdrawal at the ATM, you have to pay a _____ fee.

6. There is no _____ for transactions that are done in person inside the bank.

7. Your monthly _____ will show all the transactions of your account in one month.

8. Forty is a _____ of twenty.

B. Choose one of the following words to fill in each blank.

insufficient	invalid	incorrect

This is a difficult day for Gonzalo. He wants to withdraw thirty dollars from his account, but this is an _____ amount. He enters his PIN but finds out that he has entered an _____ number. After he corrects everything, he finds out that he doesn't have enough money in his account to withdraw the sum he needs. That's right. This time the transaction is denied because of _____ funds.

C. There is a magic word that fits all the following blanks. Can you think of the word? Write it in the blanks.

Ignacio doesn't know how much money he has in his account, and he needs to _____ a balance inquiry. He then needs to _____ a cash withdrawal from his checking account. After that, he wants to _____ a deposit and transfer funds from savings to checking. He has to decide where to _____ all of these transactions.

Focus on Grammar

A. Fill in the blank spaces with the missing forms of the verbs. The first one is done for you.

PRESENT	PAST	PAST PARTICIPLE
steal	stole	stolen
_____	_____	denied
_____	located	located
_____	_____	deducted
_____	linked	_____
_____	_____	processed

B. Fill in the blanks with words from the chart above.

1. The information is being _____ .
2. You cannot make a withdrawal at an ATM if your bank account is not _____ to that ATM.
3. The fee will be _____ from your bank account.
4. This transaction has been _____ due to insufficient funds.
5. There are ATMs _____ in banks, supermarkets, discount stores, and even airports.
6. If your ATM card is _____, you must call your bank right away.

C. Write the correct words in the blanks so that the matching sentences in parts A and B have the same meaning.

You	The bank	somebody	The machine

Part A

Part B

1. What if my card is stolen?

= What if _____ steals my card?

2. Your transaction is being processed.

= _____ is processing your transaction.

3. Each account that you access will be charged a transaction fee by Gold Rush Bank.

= _____ will charge a transaction fee for each transaction you access.

4. Quick Cash may be withdrawn from checking only.

= _____ may withdraw Quick Cash from checking only.

Focus on Information

ATM Security and Safety Precautions

Keep your ATM access cards and PIN as safe as you would keep your checks and cash. Memorize your PIN—don't write it on your card or in your checkbook. Be aware of your surroundings when using an ATM, especially at night. When you make transactions after dark, try not to go to the ATM alone.

When using an ATM, stand squarely in front of the machine to keep your transaction as private as possible. Cover your PIN entry with your hand for greater privacy. When waiting to use an ATM, please respect the privacy of those using the machine. Always take your receipt with you at the end of a transaction to assure your financial privacy. Keep your receipts and use them to check your monthly statement.

A. Discuss the following questions with a partner or a group.

1. Why should you keep your ATM access cards and PIN as safe as your checks and your cash?
2. Why should you try to take someone with you when using an ATM at night?
3. Which is safer, making transactions at the bank or at the ATM? Which is cheaper?
4. ATMs have become necessary for many people. Why do you think that is?
5. What are some advantages and disadvantages of using ATMs in your community?

B. Match the meanings on the right with the words on the left. Write the letter of the meaning in the blank next to the word it matches.

Words	Meanings
_____ security	a. at night
_____ precautions	b. safety
_____ be aware	c. personal, secret
_____ surroundings	d. facing straight forward
_____ after dark	e. ways to be careful
_____ squarely	f. the area around you
_____ private	g. pay attention
_____ privacy	h. area or information for one person only

Coping Skills

When Gonzalo's friend told him about getting cash at an ATM, he wanted to learn how to use one himself. He wanted information about all the services available at ATM machines and about safety and privacy when using them.

Put a ✓ next to the steps he took to find out everything he wanted to know about ATMs.

( ) asked friends for information

( ) spoke to a bank representative

( ) asked questions about possible problems

( ) asked about available services

( ) got more information from a brochure

( ) read the instructions on the message screen

( ) followed the written instructions for steps in a procedure

( ) followed written instructions for correcting mistakes

With a group, discuss the coping skills above.

What is the most important skill you learned in this unit? Which ones will you use in the future? In what situations will you use them? What are some things that you would do in Gonzalo's situation?

Community Assignment

1. Work in groups. Assign members of each group to go to different banks to get information on accounts, fees, ATM services, and so forth. Read and discuss brochures or other information in the group. Evaluate the costs at each bank and decide which bank offers the best services for your needs.

2. Apply for an ATM card from your bank. Use it for one transaction and share the experience with your class. Bring in the receipt. Share the type of transaction and the steps you took. Do not share your PIN!

Was That a Hamburger Combo?

Do you like to go to fast-food restaurants? Do you ever use the drive-thru window? Have you ever worked in a fast-food restaurant?

Finding a Part-Time Job

Greg Parker is a high-school senior. He wants to find a part-time job so that he can start working now and get some work experience before he graduates. His idea is that if he begins now he will be able to move up to better and better jobs quickly. He wants to apply for a job at one of the fast-food restaurants in his neighborhood. They have good work schedules for students, and people with no work experience can usually get jobs there.

1. Why does Greg Parker want to find a new job?
2. Where does he want to apply for a job?
3. Why does Greg want to apply for a job there?
4. When does he want to start working?

Look and Listen 👁 👁 👂

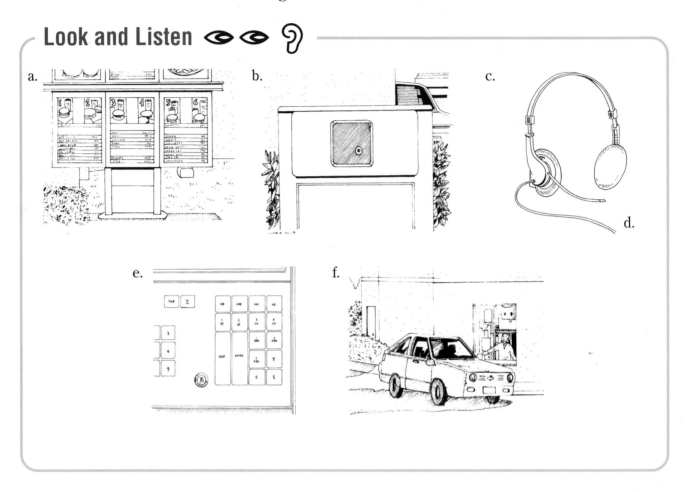

a.

b.

c.

d.

e.

f.

A. Listen to these sentences about the pictures and find the letter that matches the description. Write the letter in the blank.

1. You will be using this cash register here. _____
2. The customers look at this and decide what to order. _____
3. When customers come to the drive-thru window, you take their money and give them the change. _____
4. Customers place their orders through the speaker out there. _____
5. You will wear this headset to take orders. _____
6. Speak clearly into the microphone. _____

B. Match the items to the picture. Write the correct letter in the blank.

_____ headset and microphone   _____ cash register   _____ speaker

_____ menu   _____ drive-thru window

Focus on Listening 1

When You Use the Mike, Don't Mumble

Words	Meanings
mumble	speak quickly and quietly in a way that is hard to hear and understand
clarification	getting more information in order to understand better
punch in	key in; enter a number
hand	give something in your hand to someone

Guess what! Greg got the job. He was just hired at his neighborhood Burger Stop and will work there on Tuesdays and Thursdays after school. This is his first day on the job. The manager, Joe, is giving him instructions on some of the things he is supposed to do.

A. Listen to the conversation and check true (T) or false (F).

		T	F
1.	Greg will serve the customers inside.	_____	_____
2.	He will take orders and take money.	_____	_____
3.	Joe wants him to mumble into the mike.	_____	_____
4.	He will give change to the customers.	_____	_____

B. Answer the following questions.

1. What are Greg's job duties?

2. What is it that Joe tells him not to do?

3. Why does Joe advise Greg to repeat the order?

C. Listen to the following sentences and fill in the blanks with *If, When, After,* or *As.*

1. _____ you use the mike, don't mumble.

2. _____ you don't understand or hear an order clearly, check it again.

3. _____ you hear each order, punch it in.

4. _____ you get the whole order, tell the customer the total and direct him or her to the first window.

5. _____ customers drive to the window, take their money and give back the change.

Acknowledging Instructions

When you are learning a new job, it's very important to communicate well with your manager, supervisor, or other person who is training you. When someone is giving you instructions, you need to indicate that you are listening by saying "I see," "I understand," or "uh-huh" as the person is speaking. You also need to acknowledge the instructions. *Acknowledge* means that you let the person know that you will follow the instructions. For example, after you hear an instruction, you can say, "Okay, I will" if it is something that you should do and "Okay, I won't" if it is something that you should not do.

A. Read the instructions that Joe gave Greg. Mark each instruction with a (+) if it is positive (something that Greg should do) and a (–) if it is negative (something that Greg shouldn't do).

_____ Don't mumble in the microphone.   _____ Speak slowly and clearly.

_____ Take the customers' orders.   _____ Key in the order on the cash register.

_____ Don't speak too quickly.   _____ Repeat the whole order.

_____ If you're not sure, check the order   _____ Tell the customer the total and take the
      again.   money.

_____ Don't forget the change.   _____ Direct customers to the next window.

B. Work with a partner. Practice giving and acknowledging instructions. The first person gives one of the instructions above. The second person responds to indicate understanding. Then switch roles.

Examples: **Student A:**   Take the customers' orders.

Speak slowly and clearly.

Don't mumble in the microphone.

Don't forget the change.

Student B: Uh-huh. / I understand. / I see. / Okay, I won't.

What Should I Say?

Joe, the manager, is explaining to Greg how to take orders at Burger Stop. The Burger Stop menu has hamburgers, cheeseburgers, french fries, and drinks. There is also a combo that includes a hamburger or cheeseburger, medium fries, and a medium drink. If the customer wants a large order of fries and large drink with the combo, it costs thirty-nine cents more. That's called an *extra-size combo*.

A. Read the conversation below as you listen to the tape.

Joe: Now I want to show you how to take orders at the drive-thru.

Greg: Okay. What exactly should I say?

Joe: First say, "Are you ready to order?" or "May I take your order?"

Greg: Okay.

Joe: Then ask what they'd like to drink with that.

Greg: What would you like to drink with that?

Joe: Right. If they order a combo, ask if they want to extra-size it.

Greg: So I can ask, "Would you like to extra-size it?"

Joe: Exactly. Then repeat the order back to the customer and tell him or her the price. Say, "That'll be $3.89 at the first window, please."

B. Read the conversation above again and underline the things Greg should say to each customer. Then fill in the blanks to complete Greg's lines in the conversation below.

Greg: Welcome to Burger Stop. May I _____?

Customer: Yes. I'll have a cheeseburger combo.

Greg: _____ extra-size it?

Customer: Yeah, sure.

Greg: All right. What _____?

Customer: Diet cola.

Greg: One cheeseburger combo with diet cola. _____

_____, please.

Customer: $3.89?

Greg: Yes. At the first window, please.

C. Work with a partner. Practice the conversation between Greg and the customer. Then switch roles.

This is the menu at Burger Stop, where Greg is starting his new job. His manager said that one of the first things he needs to do is study the menu and memorize it.

A. Read the menu below.

Welcome to Burger Stop

Sandwiches

Hamburger	$ 1.75
Cheeseburger	2.25
Bacon Cheeseburger	2.75
Grilled Chicken	3.00

Beverages

Sodas:

Small	$ .79
Medium	.99
Large	1.15
Extra Large	1.35

Cola, Diet Cola, Root Beer, Orange, Lemon-Lime

Shakes	$ 1.19

Vanilla or Chocolate

Tea or Coffee	.55
1% Milk	.99
Orange Juice	1.19

Sandwich Combos
Includes medium fries and medium soft drink

#1 Hamburger Combo	$ 3.00
#2 Cheeseburger Combo	3.50
#3 Bacon Cheese Combo	3.75
#4 Grilled Chicken Combo	4.00

EXTRA-SIZE IT!

Just say **"EXTRA-SIZE IT"** and get an EXTRA LARGE FRIES AND DRINK with your COMBO!

Just add 39¢

Side Choices

Chicken Strips, 8 pc.	$ 3.00

Fries

Small	$ .95
Medium	1.10
Large	1.25
Extra Large	1.50

Onion Rings

Small	$ .95
Large	1.25
Green Salad	2.75
Desserts	1.25

Apple Pie
Ice Cream Sundae
Chocolate Chip Cookies

Food for Kids
Includes small fries, small drink, and toy

Chicken Strips, 6 pc.	$ 2.75
Hamburger	1.99
Cheeseburger	2.29

with shake: 50¢ extra

B. Read the sentences below and check true (T) or false (F). Check the menu if you are not sure.

		T	F
1.	A combo comes with fries and a drink.	_____	_____
2.	You can extra-size anything on the menu.	_____	_____
3.	You can get one order of extra large fries and a large drink for only thirty-nine cents.	_____	_____
4.	Ice cream is more expensive than apple pie.	_____	_____
5.	A hamburger meal for kids with a shake costs $2.29.	_____	_____
6.	A Grilled Chicken is a sandwich.	_____	_____

C. Read the questions below and circle the letter of the best answer. Check the menu if you are not sure.

1. Fries will not be included with an order of

   a. a combo   b. a kid's meal   c. chicken strips   d. a sandwich

2. You can have a hamburger for

   a. $3.00   b. $1.99   c. $1.75   d. all of the above

3. How much would an extra-size #4 combo cost?

   a. $3.39   b. .39   c. $4.39   d. $4.14

4. How much would a chicken strip combo with onion rings cost?

   a. $4.20   b. $3.95   c. $5.25   d. There is no chicken strip combo.

5. What comes with a sandwich combo?

   _____

6. What comes with a "Food for Kids" order?

   _____

Welcome to Burger Stop. May I Take Your Order?

Greg has started working at the drive-thru window regularly. He has many customers every day, especially from 5 P.M. to 7:30 P.M. He always tries to be polite to the customers and to make sure he repeats the order so that he won't make any mistakes.

A. Listen to the first customer's order on the tape and look at the chart below. Check to make sure that Greg wrote down the order correctly.

Item	Customer #1	Customer #2	Customer #3	Customer #4	Customer #5
Hamburger Combo C DC RB O L					
Cheeseburger Combo C DC RB O L	1 RB				
Bacon Cheese Combo C DC RB O L					
Grilled Chicken Combo C DC RB O L					
Hamburger					
Cheeseburger	1				
Bacon Cheeseburger					
Grilled Chicken					
Chicken Strips					
French Fries S M L XL	1 L				
Onion Rings S L					
Green Salad					
Kid's Burger					
Kid's Cheese					
Kid's Chicken					
Kid's Shake - Vanilla Chocolate					
Cola S M L XL					
Diet Cola S M L XL					
Root Beer S M L XL					
Orange S M L XL					
Lemon-Lime S M L XL					
Vanilla Shake					
Chocolate Shake					
Apple Pie	1				
Ice Cream Sundae					
Cookies					

B. Listen to the next four customers give Greg their orders. Mark the chart for each order that you hear. Listen as many times as necessary.

Focus on Information

Making Change

When you use a cash register, you type in the amount of money the customer gives you, and the amount of change will appear on the screen. You tell the customer this amount, and count it aloud. For example, if the purchase is $5.50 and the customer gives you twenty dollars, the screen will say $14.50. You say, "Your change is $14.50," and you hand the customer a ten-dollar bill, four one-dollar bills, and fifty cents, saying, "ten, eleven, twelve, thirteen, fourteen, and fifty cents."

Example: Greg: Your total is $12.29.
            Customer: Here's fifteen dollars.
            Greg: Okay. Your change is $2.71—two, twenty-five, fifty, sixty, seventy, and 1.

Practice making change with a partner. Use these amounts:

1. The cost is $8.50. The customer hands you a twenty-dollar bill.

2. The cost is $4.89. The customer hands you a ten-dollar bill.

3. The cost is $ 15.45. The customer hands you a ten-dollar bill, a five-dollar bill, and a one-dollar bill.

4. The cost is $11.70. The customer hands you two ten-dollar bills.

5. The cost is $6.55. The customer hands you a ten-dollar bill and five cents.

Focus on Listening 3

Checking Customers' Orders

There are several questions that Greg can use to be sure that he understands the customers' orders. Here are some expressions he can use:

What was that again?
Was that a _____?
A _____ and a what?
Did you say _____?
Could you repeat that, please?
Could you speak up, please?

A. Listen to Greg's conversations with customers. Put a check mark next to the expressions above that you hear him use.

B. Listen to the four conversations again. For each conversation, write down the expression that Greg uses to understand the customer's order.

1. _____

2. _____

3. _____

4. _____

Focus on Grammar

Change the following sentences into questions and write them on the lines.

1. You would like something to drink with that.

_____

2. You want fries or onion rings.

_____

3. You will have a dessert with your order.

_____

4. That will be all.

_____

5. You would like a large soda.

_____

6. He wants a cheeseburger combo.

_____

7. You are ready to order.

_____

8. You would like to make it extra-size.

_____

Focus on Reading

One Manager's Opinion

Words	Meanings
offended	upset
(be) in the habit	to do usually, without thinking about it
(be) likely to	a person will probably do it
take seriously	to be serious about something

A. Listen to the tape as you read what Chuck King, manager of a Better Burger restaurant in southern California, has to say about working in a fast-food restaurant.

I think the most important thing for nonnative English speakers to learn when they start working the drive-thru speaker is to speak slowly and clearly and not get nervous. That's the real problem: a lot of people just get nervous about speaking, and then it's much harder for them—and for the customer!

Another thing is that they really have to speak English at all times, especially when they're in front of customers. A lot of customers are offended when they hear employees speaking in another language. If they're talking in the kitchen, it's not so bad, but they really need to be in the habit of always speaking English when they're working. That way, they won't be as likely to forget and speak their other language at the wrong time.

I prefer to hire people from other countries to work here. They usually take their jobs more seriously and work harder than some of our native people do. When I interview someone from another country who is really interested in the job, who wants to learn and is ready to work hard, I know I've found a good worker.

B. Discuss the following questions.

1. What are three suggestions Mr. King gives for communicating through a drive-thru microphone?

2. What does he say about speaking in your native language at the restaurant?

3. Is it okay to speak your native language with other workers who speak the same language in the kitchen?

4. What is the most important thing Mr. King looks for in an employee?

5. Why should he hire you to work in his restaurant?

Role Play

Work with a partner. One person is the customer, and the other is the cashier. Practice ordering, taking orders, paying, and making change with your partner. Use the menu on Page 48. After each practice, switch roles. Don't forget to use the expressions you learned on Page 51 to make sure you understand the order. And don't forget—the customer is always right!

Coping Skills

When Greg started his new job at Burger Stop, he had many things to learn. What did he do to understand and remember everything?

Put a ✓ next to the things he did to learn his new job well. Give examples.

( )   followed oral instructions

( )   checked oral instructions to make sure he understood

( )   acknowledged oral instructions

( )   repeated customers' orders to make sure they were correct

( )   asked for repetition if something was not clear

( )   practiced what he would say to customers

( )   learned prices of menu items

( )   practiced making change

With a group, discuss the coping skills above.

What is the most important skill you learned in this unit? Which ones will you use in the future? In what situations will you use them?

Community Assignment

1. If you are looking for a job now, go to a fast-food restaurant and ask for an application. If the manager is there, ask when you can come back to speak to him or her about future job openings at the restaurant. (Be sure you dress nicely for this assignment, and go to the restaurant alone— not with friends or family.)

2. Work with a group. Members of each group go to different fast-food restaurants. Look at the menu, order food if you wish, and listen to the customer service workers as they do their job. Report to your group on the selection of food, prices, and quality of customer service at each restaurant.

Twenty on Eight, Please

Have you ever worked in a store? What kind of store was it? Did you use a cash register? Did you enjoy your job?

Hi. How can I help you?

Where would you hear these conversations?

	Customer #1	Customer #2	Customer #3
Gus:	Hi	Hi. How are you?	How can I help you?
Customer:	Hi. Twenty on eight.	Great. Give me ten dollars on two, please.	Fifteen dollars on six, and this, please.
Gus:	Okay. Thank you.	Sure. Ten dollars on two. Have a nice day.	That'll be fifteen twenty-five. Thank you.
Customer:	Thank you.	Thanks. You, too.	You're welcome. Thank you.

Answer these questions about the conversations.

1. What's the cashier's name?
2. Where does he work?
3. What are the customers buying?
4. What information are they giving the cashier?
5. How are the customers paying for their gas?

Focus on Reading 1

Vahik's New Job

Words	Meanings
equipment	machines, tools, and so forth
high-tech	short for high technology; something that has computerized equipment and is very modern in style
merchandise	items for sale in a store
display case	glass cabinets for showing merchandise
keep track of	remember many things; pay attention to many things at once
owner	person who buys a business, building, car, and so forth
boss	person who supervises you or your group of workers on the job

Vahik has just gotten his first job in the United States. He's going to be a cashier in a gas station. He's a little nervous, because this gas station doesn't look like the ones in his country at all! The equipment is very high-tech in this place. The cash register looks complicated, and there are shiny new display cases with lots of merchandise for sale.

Outside there are sixteen gas pumps. Customers can pay at the pump with ATM cards and credit cards, or they can come inside and pay for gas and other items at the counter. It looks so complicated, Vahik is not sure how the cashier can keep track of everything.

Vahik's boss is Gus, the owner of the station. Gus is an immigrant from Lebanon. He left his home country in 1980 and moved to Los Angeles, California. His first job was in a gas station, too. He worked in different gas stations for many years and learned all about the business. After learning as much as he could, and saving as much money as he could, he finally opened his own gas station three years ago. Business is great, and his station is always busy—twenty-four hours a day.

Read the story and check true (T) or false (F).

		T	F
1.	Vahik has worked in an American gas station before.	_____	_____
2.	This gas station looks like the ones in his country.	_____	_____
3.	Vahik is nervous about his new job.	_____	_____
4.	Gus has worked in gas stations for a long time.	_____	_____
5.	Business is good at Gus's gas station.	_____	_____

Focus on Listening 1

This Is a Busy Place!

The first time Vahik went to Gus's gas station, he was really surprised. There were so many customers coming in and out! Gus was working on the cash register, and he was helping a new customer every few minutes. Besides that, there were credit card customers who paid at the pump outside, and there were others who came in to buy milk, candy, food, and drinks in addition to their gas. Some customers paid by check or ATM card, but most of them used credit cards or paid in cash.

Read the chart below. Then listen to the conversations between Gus and four customers and fill in the chart. Listen as many times as necessary to get the information.

	How much?	Which pump?	Cash or credit?
First customer			
Second customer			
Third customer			
Fourth customer			

Following Directions

This is Vahik's first day of work. He is watching and listening as Gus helps the customers, and he's learning how to use the cash register. Gus is explaining everything.

A. Listen to the conversation between Vahik and Gus and answer the following questions.

1. What day is it?

2. What is Vahik doing?

3. What is Gus showing him?

4. Is it complicated?

5. What does *ring up* mean?

B. Listen to the conversation again, and look at the keypad below. Circle each button that Vahik needs to press to ring up the sales.

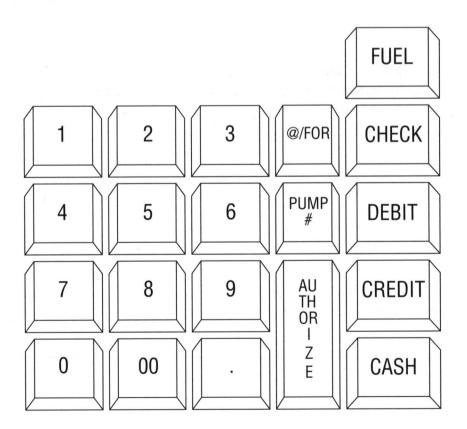

C. Look at the buttons you circled on the keypad in Exercise B. Do you remember in what order they were pressed, that is, which button was first, second, third, and so on? Write down the buttons in order here.

First Customer

Second Customer

D. Now listen to the next eight customers. All are paying with cash. Write down the buttons you need to press for each cash sale. Number one is filled in for you.

1.	8	PUMP #	2	0	00	FUEL	CASH
2.							
3.							
4.							
5.							
6.							
7.							
8.							

How Do I Do a Credit Card Sale?

After he learned how to do a cash sale, Vahik asked Gus how to do a credit card sale.

A. Listen to the conversation between Vahik and Gus and check true (T) or false (F). Listen as many times as necessary.

	T	F
1. Vahik is asking Gus about turning on one of the gas pumps.	___	___
2. They are talking about helping the customers outside.	___	___
3. The customer sometimes brings a credit card inside before pumping gas.	___	___
4. Vahik knows how to swipe the credit card.	___	___
5. The customer will sign the receipt.	___	___

B. Answer the following questions.

1. Can the cashier turn on the pumps? How?
2. What does *fill up* mean?
3. What does *authorize* mean?
4. What does *print out* mean?
5. What were the steps for a credit card sale?
6. How did Vahik check his understanding of the steps for a credit card sale?
7. What buttons on the cash register did Vahik learn to use in this conversation with Gus?

C. Listen to Vahik's questions and fill in the blanks. Listen as many times as necessary.

1.

Vahik: Gus, can I _____ you something?

Gus: Sure. What _____ it?

Vahik: Well, I noticed that _____ you do a credit card sale, _____ you take the customer's _____ and put it over there. _____ is that?

Gus: Well, that's for people who want to _____ their tank.

2.

Vahik: You _____ you turn on the pump for the customer?

Gus: Yes. I do it on the _____, right here.

3.

Vahik: How _____ I ring up the sale?

Gus: Well, _____ press PUMP #, then CREDIT, and then _____ swipe the card.

4.

Vahik: Do _____ swipe the card right here?

Gus: Yes. There will be a _____ on the _____ that tells you when to do it.

5.

Vahik: And _____ I ask the customer to _____ after that?

Gus: Yes, that's _____. You wait for the _____ to print out, and then _____ ask the customer to sign it.

6.

Vahik: Which copy _____ I give the _____?

Gus: The _____ copy.

D. Work with a partner. The first person asks how to do something at work, and the second person explains. Then the first person repeats the steps, and the second person listens to see if the steps are right.

STUDENT A

How do I . . .

    ring up a cash sale?
    do a credit card sale?
    turn on the pump?
    keep track of the
       customers' credit cards?

Let me see if I've got that right. I . . .,
and then I . . . (repeat the steps)

Okay, thanks a lot.

STUDENT B

Well, you . . . (explain the steps)

Yes, that's right.
    (or)
No, you need to . . ., and then you need to . . .

What Are These Keys For?

Read the questions for each part of the cash register pictured below. Then listen to the conversation, look at the keys, and answer the questions. Listen as many times as necessary for each picture.

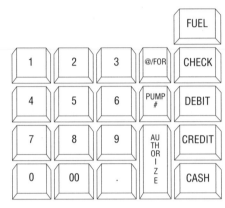

A.

1. Where are these keys on the keyboard?

2. What are they for?

3. Will Vahik use these keys often?

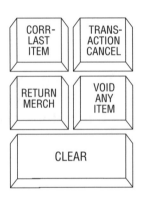

B.

1. Are these keys important?

2. What are they for?

3. What does CORR stand for?

4. What does TRANSACTION mean?

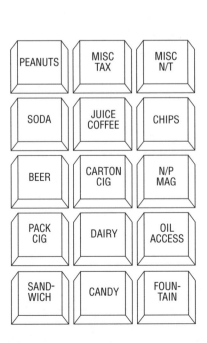

C.

1. What are these keys for?

2. What does *FOUNTAIN* mean?

3. What does *CIG* stand for?

4. What does Gus say is the most important thing?

D.

1. What are these keys for?

2. Does Vahik need to use them?

3. What did Vahik say when he didn't understand?

4. What does EOD mean?

5. Which key means end of shift?

6. Did Vahik understand the word *safe?*

7. What did he say to clarify the meaning?

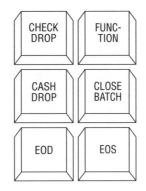

1	2	3	4	5	6	7	8
9	10	11	12	13	14	15	16

1 Item	Fuel Pre -pay		20.00
	SUB TOTAL		20.00
	TAX		0.00
	TOTAL		20.00
	Cash		20.00
CHANGE	Cash		0.00
RECEIVED	20.00		0.00

E.

1. What are the two screens for?

2. What does the screen for the gas pumps show?

3. What does the screen for the current transaction show?

4. What did Gus say to Vahik at the end of the conversation?

5. How did Vahik answer him?

It Looks So Complicated

This is the cash register at Gus's gas station. There are two screens at the top, a large keyboard with many keys, and a cash drawer under the keyboard. Also, there is a slot next to the keyboard on the right side for swiping a credit card or ATM card. This electronic cash register can calculate, keep track of transactions, keep track of the gas pumps outside, and help the cashier make change. It can do everything except make coffee in the morning!

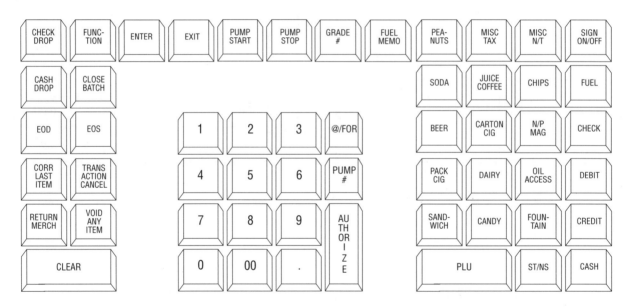

1	2	3	4	5	6	7	8
9	10	11	12	13	14	15	16

1 Item	Fuel Pre -pay	20.00
	SUB TOTAL	20.00
	TAX	0.00
	TOTAL	20.00
	Cash	20.00
CHANGE	Cash	0.00
RECEIVED	20.00	0.00

A. Read the cash register screens and keyboard above and check true (T) or false (F).

		T	F
1.	There are sixteen spaces on the gas pump display window.	_____	_____
2.	The buttons for entering the kind of transaction are on the left.	_____	_____
3.	The PUMP # key is next to the numbers 1 through 9.	_____	_____
4.	The keys for different kinds of merchandise are grouped together.	_____	_____
5.	The keys for making corrections are in the bottom right corner.	_____	_____
6.	The current customer needs change for a twenty-dollar bill.	_____	_____

B. Read the abbreviations on the keys of the cash register and match them with their meanings below. Write the abbreviations in the blanks.

_____ = oil and car accessories (things for cars)

_____ = newspapers and magazines

_____ = miscellaneous taxable merchandise (extra things that have a tax, like a toy, for example)

_____ = miscellaneous nontaxable merchandise (extra things that don't have a tax, like food)

_____ = returned merchandise (things that customers return)

_____ = correct last item

_____ = subtotal (for adding the total before the sale is finished)

_____ = no sale (for opening the cash drawer when there's no sale)

Focus on Communication 2

What Does N/T Stand For?
Use the definitions for each key in Exercise B above to give your partner information.

Student A points to the keys and asks questions about the meaning. Student B explains the meaning of each key. Then switch roles.

Examples:

Student A:	What does this key mean? (Pointing to the CORR LAST ITEM key)
Student B:	It means correct last item.
Student A:	What does N/P Mag stand for?
Student B:	It stands for newspapers and magazines.

A. Ask and answer questions about what the cash register keys are used for. Read your part silently first; then work with a partner. The first person asks the question. The second person finds the explanation on the list and says the answer. Then switch roles. The answers are not in order.

STUDENT A	STUDENT B
What is the PLU key for?	AUTHORIZE? That's for turning on the pump for a credit card customer.
What is the FUNCTION key for?	We don't use the FUNCTION key. That's for the manager only.
What's the AUTHORIZE key for?	I'm not sure about the PLU key. I think it's for special codes.
What's the SIGN ON/OFF key for?	They're for entering information in the special functions like CASH DROP and CLOSE BATCH. We don't have to use them.
What are the ENTER and EXIT keys for?	Well, SUBTOTAL is for adding up the sale before it's finished, and NO SALE is for opening the cash drawer when there's no sale.
What is the GRADE # key for?	That's the button you use to sign on at the beginning of your work day and sign off before you go home.
What is the FUEL MEMO key for?	Oh, those buttons are for starting or stopping one pump if there's a problem.
What are the PUMP START and PUMP STOP keys for?	The GRADE # button is for selecting the octane of the gas, but the customer can choose the octane outside at the pump.
What is the ST/NS key for?	I'm not sure what FUEL MEMO means. I think it prints a receipt for gas only.

B. Ask and answer questions again for keys that you're not sure about or don't remember. Switch roles for each question.

Examples: **Student A:**   What does ST mean again?
             Student B: It means subtotal.

             Student A: What was the AUTHORIZE key for again?
             Student B: It's for turning on the pump for a credit card sale.

Focus on Reading 2

First Day of Work

Whenever anyone starts a new job, it's difficult to learn what to do. There are always new ideas and new words to learn, even for native speakers of English. Everyone feels nervous on their first day of work. The important thing is to be calm, ask questions, and learn as much as you can.

One way to be sure you learn as quickly as possible is to take notes. Take a small notebook to work with you the first few days. When you have a free moment, or during your break or lunch, write down some notes. For example, the abbreviations on the keys of the cash register were new for Vahik, and some of them were hard to remember. He could write those down in his notebook, along with the meaning of each one. Then after work he could study those new words and abbreviations to get ready for the next day.

Also, when you learn a new procedure, it's a good idea to write down the steps. That will help you remember all the steps when you're working by yourself. Some procedures for the cashier in the gas station were how to do a cash sale, how to do a credit card sale, and how to do a cash drop. Try to learn the most common procedures first, and your job will be easier. For example, Vahik learned how to do a cash sale and a credit card sale before Gus taught him how to do a cash drop.

Another thing that is very important to your boss is that you learn to ask questions when you are not sure how to do something on your new job. Every boss says that the most important thing is to ask questions. That's because your boss wants you to learn your job well and do everything the way you are taught. Then you can work independently, and your boss doesn't have to worry about your work. Your boss will know that you will always check your instructions and make sure you are doing your job the right way.

A. Read the selection again. Underline the important ideas.

B. Look at all the ideas you underlined. Choose five that are the most important to you. Write notes or sentences below for five ideas you want to remember.

1. _____

2. _____

3. _____

4. _____

5. _____

Role Play

Find an item in your classroom, such as a calculator, umbrella, or projector, and explain to another person how to use the item, what the parts of the item are for, what the keys or buttons on the item are for, what the name of the item means, and any other information you can explain. The other person will ask what the parts of the item are for and will ask *How do I . . .?* questions about how to use the item. Then switch items with two other students.

The next day, bring an item or a piece of equipment to class. Explain what the piece of equipment is for, and let the other students ask you questions about it.

Coping Skills

When Vahik started working at Gus's gas station, he had to learn how to use the cash register, a complicated piece of equipment. What did Vahik do to learn his new job well?

Put a ✓ next to the things he did to learn how to use the cash register. Give examples.

( )   asked about procedures

( )   repeated the steps in a procedure to confirm them

( )   practiced new procedures

( )   learned new words

( )   asked about parts of the cash register

( )   asked about meaning

( )   learned abbreviations

( )   asked about functions

With a group, discuss the coping skills above.

What is the most important skill you learned in this unit? Which ones will you use in the future? In what situations will you use them?

Community Assignment

Go to a fast-food restaurant, store, or gas station and ask about the cash register keyboard. Ask the cashier if it was difficult to learn to use it. Come back and tell the class what you learned about that keyboard. Describe it to your group and discuss how it is different from or the same as Gus's cash

UNIT 6

Self-Service at the Post Office

Where do you usually mail packages? What time of day do you usually do it? Is it convenient? Would you like to find a more convenient way?

Busy, Busy, Busy

Natasha is talking to Vera at the office. She has a small package that she has to send to her niece in New York. It's a birthday present, and she wants it to arrive before Friday. This is a busy time at the post office, and she is a busy person. She does not have the time to wait in long lines at the post office. She usually keeps stamps at home for her regular mail, but this package is heavier and she does not know how many stamps to use.

Check true (T) or false (F).

	T	F
1. Vera is Natasha's daughter.	_____	_____
2. Natasha has a large envelope she needs to mail.	_____	_____
3. She is going to ask her husband to mail it for her.	_____	_____
4. Vera is going to help her.	_____	_____

Focus on Communication 1

I Wouldn't Do That

Natasha and Vera are talking about Natasha's mail. They are trying to find an efficient way to send the package to Natasha's niece without spending too much time at the post office.

A. Listen to the tape and then answer these questions. Listen as many times as necessary.

1. Why can't Natasha go to the post office early in the morning?
2. Why doesn't she want to go during her lunch break?
3. Why doesn't she want to ask her husband to mail the package?
4. How can she weigh her package without standing in line?
5. How can she buy stamps without waiting in line?

B. Listen to the tape again and fill in the blanks.

Vera: Maybe you _____ go before work tomorrow.

Natasha: Well, it doesn't open until 8:00 and that would make me late for work.

Vera: _____ you ask your husband to mail it for you? He's not busy.

Natasha: No, I _____ ask him. He still doesn't feel confident enough about his English to use the post office. But maybe I _____ try going after work.

Vera: No, I _____ do that. It's better to avoid the post office at that time of day. Wait! I know! You _____ use the self-service scale and the stamp machine!

Natasha: What?

Vera: There's a scale there in the lobby of the post office. Most people don't use it, but it's easy. You just put your package on it, punch in the zip code, and it tells you the postage. Then, you can buy the stamps you need at the vending machine.

Natasha: Gee. I didn't know about that. Is it hard to do?

Vera: You know what? I _____ go with you tomorrow at lunch.

Natasha: That _____ be great! Thanks!

C. Now practice the dialogue with another student.

Look and Listen 👁 👁 👂

Natasha and Vera are at the post office. Vera is showing Natasha the self-service scale and explaining how it works. Read the steps below. Then listen for the order of the steps. Number the steps from 1 to 5 in the order that they appear on the message screen. Listen as many times as necessary.

_____ Enter destination zip code.

_____ Press YES to continue.

_____ Press any key.

_____ Press NO to enter correct zip code.

_____ Place item on scale.

Real-Life Reading 1

Words	Meanings
weight	how heavy the item is (in the United States weight is measured in pounds, or lbs., and ounces, or oz.)
postage	how much it costs to mail an item
surcharge	extra postage cost for items that are too large or the wrong shape
rate comparison	a list of different prices for different mail services
information slip	a small piece of paper with information on it
domestic mail	mail that will be delivered in the United States; not international mail

A. Read the instructions on the screens carefully. Then answer the questions on the next page.

PRESS

ANY

KEY

WELCOME TO CRESCENT STATION

PLACE ITEM ON SCALE

FOR WEIGHT AND POSTAGE

ENTER DESTINATION ZIP CODE

90011

Press YES to continue.
Press No to enter zip code.

ENTER DESTINATION ZIP CODE

90012

Press Yes to continue.
Press No to enter correct zip code.

DOMESTIC MAIL INFORMATION

FIRST-CLASS POSTAGE   BASIC RATE
$1.43

Surcharges may apply to oversize or odd-size items.
Press Yes for surcharge or mailing instructions.

Answer the following questions about the information on the previous page.

1. What message was on the screen when Vera and Natasha started?
2. Where did she mail the package?
3. How do you get the weight and postage of a package?
4. What mistake did Vera make? How did she correct her mistake?
5. How much is first-class postage for Natasha's package?
6. If Natasha wants to find out if there is any surcharge for her package, what should she do?

Focus on Listening 1

How Should I Send It?

The next screen that Natasha and Vera saw on the scale looked like this. Natasha needed to decide how to send her package. She needed it to arrive at her niece's house in time for her birthday.

```
                          RATE COMPARISON

   1.  EXPRESS MAIL      SECOND DAY GUARANTEED        $11.75
   2.  PRIORITY MAIL     TWO DAYS NOT GUARANTEED        3.20
   3.  FIRST-CLASS MAIL  VARIES BY DESTINATION          1.43
   4.  THIRD-CLASS MAIL  VARIES BY DESTINATION          1.43

   ENTER SELECTION NUMBER FOR SPECIAL SERVICES.
   PRESS NO TO EXIT SCREEN.
```

Listen to Natasha and Vera's conversation and answer the following questions.

1. Which type of mail service does Natasha choose: Express Mail or Priority Mail?
2. How much will it cost to mail her package?
3. When will the package arrive in New York?
4. Will there be a surcharge? Why or why not?
5. Where will she get a Priority Mail label?
6. Where will she get a Priority Mail stamp?
7. What does *guaranteed* mean?
8. What do you think *varies by destination* means?

Natasha saw this poster at the post office. It was on the wall near the scale and the stamp machine.

Postal Rates and Fees

EXPRESS MAIL

Express Mail is our fastest service. Next day delivery by 12 noon to most destinations. Delivered 365 days a year with no extra charge for Saturday, Sunday, or holiday delivery. All packages must use an Express Mail label. Items may weigh up to 70 pounds and measure up to 108 inches in combined length and girth. Call 1-800-222-1811 for delivery information between ZIP Codes.

Features—Express Mail envelopes, labels, and boxes are available at no additional charge at post offices or by calling 1-800-222-1811.

Post Office to Addressee Service

Up to 8 ounces	$11.75
Over 8 ounces up to 2 pounds	15.75
Up to 3 pounds	18.50
Up to 4 pounds	21.25
Up to 5 pounds	24.00
Up to 6 pounds	26.75
Up to 7 pounds	29.40
Over 7 pounds (see postmaster)	

Flat Rate Envelope—Post Office to Addressee Service
$15.00, regardless of weight or destination for matter sent in a flat rate envelope provided by the Postal Service.

PRIORITY MAIL

Priority Mail offers 2-day service to most domestic destinations. Items may weigh up to 70 pounds and measure up to 108 inches in combined length and girth.

Features—Priority Mail envelopes, labels, and boxes are available at no additional charge at post offices or by calling 1-800-222-1811.

Single-Piece Rates

Up to 2 pounds	$3.20
Up to 3 pounds	4.30
Up to 4 pounds	5.40
Up to 5 pounds	6.50
Over 5 pounds (see postmaster)	

Flat Rate Envelope
$3.00, regardless of weight or destination for matter sent in a flat rate envelope provided by the Postal Service.

PARCEL POST ZONE RATES

For merchandise only.
For package rates priced by distance and weight, see postmaster.

FIRST-CLASS MAIL

Single-Piece Letter/Flat Rates

First ounce	$0.33
Each additional ounce	0.22

WEIGHT NOT OVER (OZ.)		WEIGHT NOT OVER (OZ.)	
1*	$0.33	7	$1.65
2	0.55	8	1.87
3	0.77	9	2.09
4	0.99	10	2.31
5	1.21	11	2.53
6	1.43	12	2.75
		13	2.97

SECOND-CLASS MAIL

Only authorized publishers and registered news agents may mail publications at second-class rates. The public may mail publications at the applicable Express Mail or single-piece Priority Mail, First-, Third-, or Fourth-Class rate.

THIRD-CLASS MAIL

Used primarily by retailers, catalogers, and other advertisers to promote products and services. See postmaster for details. Single-piece rates may be used by the public for mailing certain items—circulars, other printed matter, merchandise, seeds, and plants—weight less than 16 ounces.

FOURTH-CLASS MAIL

For mailing certain items—books, circulars, catalogs, other printed matter, and parcels—weighing 16 ounces or more. Enclosed or attached First-Class Mail is charged at First-Class rates. Packages may weigh up to 70 pounds and measure up to 106 inches in combined length and girth.

SIZES FOR DOMESTIC MAIL

Mail must meet these standards:
- Thickness—No less than 0.007 inch thick.
  Pieces that are 1/4-inch thick or less must be at least 3-1/2 inches high, 5 inches long, and rectangular in shape.
- Combined length and girth—No more than 108 inches.
- Weight—No more than 70 pounds.

A. Read the poster on Page 74 quickly. Match the explanations on the right with the types of mail service on the left. Write the letter of the explanation in the blank next to the type of mail service it matches.

Type of mail service

_____ Express Mail

_____ First-Class Mail

_____ Priority Mail

_____ Second-Class Mail

_____ Third-Class Mail

_____ Fourth-Class Mail

Explanation

a. for mailing books

b. used for magazines and newspapers only

c. the fastest mail service

d. used for advertising

e. delivers in two to three days

f. for most letters and postcards

B. Work with a partner. Read the following questions and decide on the best way to send each item: Express Mail, Priority Mail, or First-Class Mail.

1. You want to send a letter to your brother in another state.

2. You want to send a birthday present to your nephew in another state. His birthday is in one week.

3. Your son needs a copy of his birth certificate, and he needs it tomorrow.

4. You are applying for a job in another city. They must receive your letter of application and your resume by Thursday. It is Monday now.

5. You want to send a card and some photos to your friend. The envelope weighs five ounces.

C. Work with a partner again. Find the following words on the poster on Page 74 and guess their meaning. Then check with your teacher to see if you are correct.

delivery	length	postmaster	retailer
charge	girth	publisher	catalog
parcel	postal card	publication	cataloger
label	printed matter	news agent	advertiser

D. Work with another partner. Read the information about special services from the poster below. Then read items 1 through 4 below the poster and decide which special service or services are needed for each situation: Certificate of Mailing, Certified Mail, Insured Mail, Registered Mail, and/or Return Receipt.

SPECIAL SERVICES (DOMESTIC MAIL)

Certificate of Mailing
Proves that a mailing was mailed. Must be purchased at time of mailing. No record kept at the post office.
Fee, in addition to postage—$1.25

Certified Mail
Provides a mailing receipt, and a record is kept at the recipient's post office. A return receipt can also be purchased for an additional fee. Available only with First-Class and Priority Mail.
Fee, in addition to postage—$1.40

Insured Mail
Provides coverage against loss or damage. Coverage up to $600.00 for Third- and Fourth-Class mail as well as Third- and Fourth-Class matter mailed at Priority Mail or First-Class Mail rate. Items must not be insured for more than their value.

Liability			Fee, in addition to postage
$ 0.01	to	$ 50.00	$0.85
50.01	to	100.00	1.80
100.01	to	200.00	2.75
200.01	to	300.00	3.70
300.01	to	400.00	4.65
400.01	5o	500.00	5.60
500.01	to	600.00	6.55

Registered Mail
Provides maximum protection and security to valuables. Available only for Priority Mail and First-Class Mail rates. May be combined with COD, restricted delivery, or return receipt. Insurance up to $25.00 can be purchased by using registered mail.

| Value | | Fee, in addition to postage | |
		With postal insurance	Without postal insurance
$ 0.00 to	$ 100.00	$6.20	$6.00
100.01 to	500.00	0.00	0.00
500.01 to	1,000.00	0.00	0.00
1,000.01 to	2,000.00	0.00	0.00

For higher values (see postmaster)

Return Receipt
Available only for Express Mail, certified mail, COD, insured mail for more than $50.00, or registered mail.

Requested at time of mailing
Showing to whom (signautre) and date delivered—$1.25

Requestyed after mailing
Showing to whom (signature) and date delivered—$7.00

1. You are sending a necklace that belonged to your grandmother to your niece in another state.

2. You are sending a jacket that cost $95 to your uncle. You want to buy another one if it gets lost in the mail.

3. You are sending an important document, and you want to know that it arrived, who got it, and when.

4. You are sending payment of a bill. You want to be able to prove that you mailed it.

Now We Need a Stamp

Natasha and Vera are buying a Priority Mail stamp from the stamp vending machine.

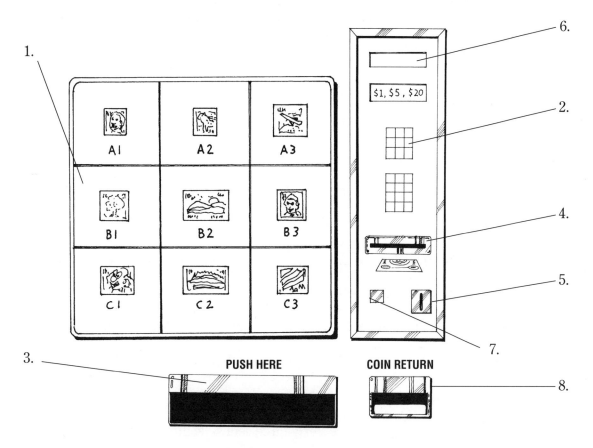

A. Listen to Natasha and Vera as they use the stamp vending machine. Then fill in the number of each item below in the blank given on the picture.

1. This is where you can choose which stamps you want to buy.

2. Make your selection here by keying in the letter and number combination for the stamps you want.

3. This is where the stamps come out after you make your selection.

4. This is where you can insert bills.

5. This is where you can insert coins.

6. This message window tells how much money you have put in the machine.

7. If you have a problem, press the button here and the machine will return your money.

8. This is where your change will come out.

B. Listen to the conversation again. Fill in the blanks.

Vera: Now we need a _____ Mail stamp.

Natasha: Here it is. Selection C2. It _____ $3.20. Let me see if I have change.

Vera: That's okay, the machine takes bills. It says right here it takes one-, five-, ten-, and

twenty-dollar bills. Here's the place to insert _____ right below the

keypad..

Natasha: Well, I have two one-dollar bills and five _____.

Vera: That'll work. You can put in the _____ here on the bottom right..

Natasha: Oh! This message screen at the top _____ me how much money I've put in!

That makes it easy.

Vera: Yeah, and if you need change, it _____ out automatically down here next to

where you pick up your stamps.

Natasha: This is fun! I think I'll buy some thirty-three-cent stamps, too. What are these

self-adhesive stamps?

Vera: Oh, I like those. You don't have to _____ them.

Natasha: Okay, this _____ "five thirty-three-cent self-adhesive stamps" for $1.65.

I'm going to get that, too. Oops, I hit B instead of C. Now what?

Vera: See that coin-return button next to the _____ you put yur coins? Try that.

Natasha: Oh, you're right. My money came back. Now I'll punch in C3 . . . more

_____ this time. There. I got my stamps and my change.

Vera: Great. We're all _____.

Natasha: Yeah. Thanks so much for _____ how to do this.

C. Check your answers with another student or with your teacher. For any new words, ask, "How do you spell _____?" or "What does _____ mean?"

Focus on Communication 2

Using a Stamp Machine, Partner 1

A. Work with a partner. The first person looks at the stamp machine selection on this page. The second person looks at the next page. Ask your partner for the information missing on your page. Write in the missing information. Your partner will ask you about information missing on the next page. Look at your information to answer.

Example:

Partner 1: What can you buy with selection A1?

Partner 2: Postal cards, a pack of five. How much are ten self-adhesive stamps?

Partner 1: $3.30. They're thirty-three cents each. How much are twenty thirty-three-cent stamps?

Partner 2: $6.60.

A1 $1.00	**Ten thirty-three-cent, self-adhesive stamps** **A2** $3.30	**Twenty thirty-three-cent self-adhesive stamps** **A3**
Pack of five forty-six-cent stamps **Canada rate for 1/2 oz.** **B1** $2.30	**B2** $2.30	**Air Mail postage** **Five sixty-cent stamps** **Letter rate for other countries** **B3** $3.00
Domestic Express Mail **Express Mail for up to 8 oz.** **C1**	**Domestic Priority Mail** **Includes stamp and label** **C2** $3.20	**Five thirty-three-cent self-adhesive stamps** **C3** $1.65
One thirty-three-cent stamp **Two one-cent stamps** **D1** .35	**Stamp booklet** **Fifteen thirty-three-cent self-adhesive stamps** **D2**	**Stamp booklet** **100 thirty-three-cent stamps** **D3** 33.00

B. Now, together with your partner, compare tables. Did you both write in the missing information correctly?

Using a Stamp Machine, Partner 2

A. Work with a partner. The first person looks at the stamp machine selection on the previous page. The second person looks at this page. Ask your partner for the information missing on your page. Write in the missing information. Your partner will ask you about information missing on the previous page. Look at your information to answer. Continue until your chart is complete.

Example:

    Partner 2: How much are ten self-adhesive stamps?

    Partner 1: $3.30. They're thirty-three cents each. How much are twenty thirty-three-cent stamps?

    Partner 2: $6.60. What can I buy with selection B1?

    Partner 1: A pack of forty-six-cent stamps.

Postal cards Pack of five twenty-cent cards A1 $1.00	Ten thirty-three-cent self-adhesive stamps A2	Twenty thirty-three-cent self-adhesive stamps A3 $6.60
B1 $2.30	Pack of five forty-six-cent stamps Mexico rate for 1/2 oz. B2 $2.30	Air Mail postage Five sixty-cent stamps Letter rate for other countries B3
Domestic Express Mail Express Mail for up to 8 oz. C1 $11.75	C2 $3.20	Five thirty-three-cent self-adhesive stamps C3 $1.65
D1 .35	Stamp booklet Fifteen thirty-three-cent self-adhesive stamps D2 $4.95	Stamp booklet 100 thirty-three-cent stamps D3

B. Now, together with your partner, compare tables. Did you both write in the missing information correctly?

Focus on Grammar

A. Study the following sentences.

1. Vera: Couldn't you ask your husband to help you?
2. Natasha: No, I wouldn't do that. He wouldn't feel confident going to the post office.
3. Vera: I wouldn't go to the post office after work. It's too crowded at that time of day.
4. Natasha: Maybe I could try going during lunch.
5. Vera: You could use the self-service scale and the stamp machine.
6. Natasha: I don't think I would know what to do.
7. Vera: I know! We could go together during lunch.
8. Natasha: That would be great!

Grammar Note:

Would is a form we use when talking about situations that we only imagine.
Example: If I had a million dollars, I would buy a small plane.

It is also used in situations when we are thinking about what would be a good thing to do.
Example: It would be a good idea to send the light package by Priority Mail.
It wouldn't be a good idea to send a heavy package by Express Mail.

Could is a form we sometimes use as a suggestion.
Example: Maybe you could ask your husband to help you.

It is also used to offer to do something.
Example: I could help you with that if you want.

B. Fill in the blanks, choosing *have to, could, can't,* or *wouldn't*.

1. You _____ send out the letter before the end of the month.
2. You have some choices. You _____ use the scale and the vending machine instead of waiting in line.
3. You _____ get a receipt to prove that you have sent the letter on time.
4. I don't think it's necessary to use Registered Mail for this letter. I know I _____.
5. I _____ go to the post office in the morning because I _____ be at work.

Vocabulary

Fill in the blanks using the appropriate word from the list below.

Domestic	rate	surcharge	destination	selection number

There is a _____ for oversize items.

The _____ for Express Mail is higher than Priority Mail.

For First-Class Mail, the delivery time varies by _____.

Enter the _____ for Priority Mail.

_____ rates are different from international rates.

Coping Skills

In all of her experiences at the post office, Natasha had to make decisions and take actions. Which of the following steps did she take? Put a ✓ next to the steps that she (or Vera) used. Give examples.

( )   asked for advice

( )   offered to help

( )   read instructions

( )   followed written instructions

( )   repeated the steps she needed to follow

( )   focused on the problem

( )   analyzed options or choices

With a group, discuss the coping skills above.

What is the most important skill you learned in this unit? Which ones will you use in the future? In what situations will you use them?

Community Assignment

1. Go to your local post office. If you have something to send, place it on the scale and find out how much postage you need to send it to the desired destination. If you do not have anything to mail, just put a book on the scale and find out the postage. Use your home zip code if you wish. Write down the weight, the amount of postage, and the type of delivery.

2. Use the stamp vending machine in the same way. Buy some stamps. Write down the number and the letter of the buttons to press. Write down the number of stamps and the amount you have to pay for your choice. What kind of money do you have to insert in the machine? How much change will you receive?
    NOTE: Try to put in a bill large enough so that your change will be more than one dollar. You will probably receive this dollar (or dollars) in an interesting form. Share your experience with the class.

Springfield Public Library

Have you ever used a library? How do you find books in a library? What is the difference between a card catalog and an electronic catalog?

Looking for a Better Job

Vinh has been in the United States for over a year now. He has worked at different part-time jobs in the evenings and attended school during the day. Now that his English is better, he wants to find a better job. The trouble is, he doesn't know much about finding a job in this country. He went to the library to try to find a book on looking for a job, but he couldn't find what he needed. He has decided to go to school and ask his teacher for help.

Read the story and check true (T) or false (F).

		T	F
1.	Vinh has been in the U.S. for over a month.	_____	_____
2.	He has attended night school.	_____	_____
3.	Vinh went to the library to apply for a job.	_____	_____
4.	He needs information on how to find a job.	_____	_____
5.	He thinks his teacher can help him.	_____	_____

Look and Listen 👁 👁 👂

A. Look at the illustration and listen to the tape.

THE
ENGLISH
ZONE, BOOK 1

*An Integrative Course in
Communicative English*

KAREN BATCHELOR

RANDI SLAUGHTER

Dominie Press, Inc.

Published by:
Dominie Press, Inc.
1949 Kellogg Avenue
Carlsbad, CA 92008 USA

ISBN 1-56270-974-7
Printed in Singapore by PH Productions Pte. Ltd.
1 2 3 4 5 6 PH 01 00 99

B. Listen to the tape. Number the sentences to match what the librarian says.

_____ The date of publication is 1999.

_____ The title is the name of the book.

_____ The publisher is Dominie Press, Inc.

_____ The author is the writer.

C. Fill in the information about the book shown above.

Author:

Title:

Date of Publication:

Publisher:

D. Look at the cover and the first two pages inside the cover of this book, *English for Technology*. Find the complete title, authors, publisher, and date of publication.

I Don't Know How to Do a Resume

Vinh is talking to his English teacher, Ms. Martin. He is asking for some advice on how to apply for jobs and how to write a resume.

Words	Meanings
resume	a list of someone's education, work experience, and qualifications; often submitted with an employment application
opening	an unfilled position
classified ads	advertisements arranged in classes, or categories, in the newspaper
search	look for something
keyword	a word or words typed in a computer to search for something
card catalog	a series of cards, arranged in alphabetical order, with information about all the books in a library
online catalog	a computerized list of information about all the books in a library

A. Listen to the conversation between Vinh and Ms. Martin and answer the following questions.

1. What is Vinh having trouble with?

2. Has he looked for any jobs yet? Has he applied for any?

3. What does he need to learn?

4. What does Ms. Martin advise him to do?

5. What kind of information can he get at the library?

6. How can he get information at the library?

7. Have you ever written a resume?

8. How could a book from the library help you when you need to look for a job?

B. Listen to the tape carefully. Fill in the blanks with the words you hear. Listen as many times as necessary.

1. Ms. Martin: _____ you ever _____ before?

   Vinh: I _____ _____ at a lot of part-time jobs since I came here.

2. Ms. Martin: _____ you _____ at the job bulletin board here at school?

   Vinh: The last time I checked, there _____ no openings listed.

3. Ms. Martin: You should also read the classified ads in the newspaper.

   Vinh: I _____ already _____ that. Two weeks ago, I found some openings in the classified ads and applied for them, but nothing _____.

4. Ms. Martin: You mean you _____ _____ anything from them yet?

   Vinh: Well, two of them answered. And they both _____ for a resume.

   Ms. Martin: _____ you _____ one?

   Vinh: That's just the problem. I don't know how to do a resume. I _____ never _____ a resume before.

5. Ms. Martin: It might be a good idea to go to the library and look for some books.

   Vinh: I _____ to the library last week. But I _____ _____ how to look for what I needed.

6. Ms. Martin: _____ you _____ the librarian for help?

   Vinh: No, because I _____ _____ the name of any books to look for.

7. Ms. Martin: You don't always need the title. You can search under "Subject" or "Author."

   Vinh: I _____ , but I _____ find anything in the card catalog.

C. Compare your answers with a partner or with the class.

D. Work with a partner. Use the following cues to form questions and answers.

Example 1:   Have you ever worked?

                          Yes, I've worked at some part-time jobs.

STUDENT A   **STUDENT B**

Have you ever . . .   Yes, I've . . .

- worked before?   No, I've never . . .
- looked for a job?
- written a resume?
- applied for a job?

Example 2:   Did you find the book?

                          No, I didn't know how to look for it.

STUDENT A   **STUDENT B**

Did you . . .

- go to the library?   Yes, I . . .
- find the book?   No, I . . .
- ask the librarian?

Focus on Information

Springfield Public Library

Words	Meanings
call number	the code number indicating the location of a book or magazine in the library
status	the present situation; for example, a book is now available or not available
check out	borrow (a book) from the library
loan, lend	take for a short time (to return)
slip	a small piece of paper
due date	date a library book must be returned
late fine	money to be paid if the book is returned later than the due date

The following information is from a brochure from the Springfield Public Library called *How to Use the Automated System.*

A. Read the library brochure and check true (T) or false (F) below.

Search Tips: The library's new online catalog can help you search for a book in a number of ways. You can search under title, author, subject, or key words in the title. The computer screen will provide you with the location of the book in the library, the call number by which to find it, and the current status of the book. The status will show whether or not the book is available at the time of the search. It will also show the due date by which the book is expected to be back if it has been checked out. The automated system will also provide specific information, such as the name of the publisher, the date of publication, and the number of editions.

Check-out: Most items are loaned for twenty-eight days. New books, children's books, and audio items are loaned for fourteen days, and magazines for seven days. Videos are rented for seven days. Materials are due on the date shown on the date-due slip. Late fines are two dollars per day for videos, and twenty cents per day for all other items.

	T	F
1. The call number is the number of books available in the library.	_____	_____
2. You can see a book's location in the library on the computer.	_____	_____
3. Only the librarian can tell you whether or not a book is available.	_____	_____
4. You can borrow videos at the library for free.		

B. Answer the following questions.

1. What are the four ways to search for a book in the online catalog?

2. What other kinds of information does the automated system provide?

3. How long can you normally keep books you check out from the library?

4. What are the types of items that you can keep for fourteen days?

5. What is a fine? How much are the late fines at the Springfield Public Library?

Real-Life Reading 1

Searching for a Book Online

Words	Meanings
location	the section of the library where a book can be found, such as fiction, nonfiction, new books, children's, oversize, and so on
fiction	a storybook
nonfiction	a book about facts; not a storybook
entry	one item in a computerized list
display	show on the computer screen
sort by year	put in order by year of publication

Vinh is at the library. He is using the online catalog to find a book that will help him write a resume. He doesn't know the name of a specific book or author. He doesn't know exactly what subject to look under, either.

A. Read the three message screens that Vinh saw on the computer. Answer the questions that follow each screen.

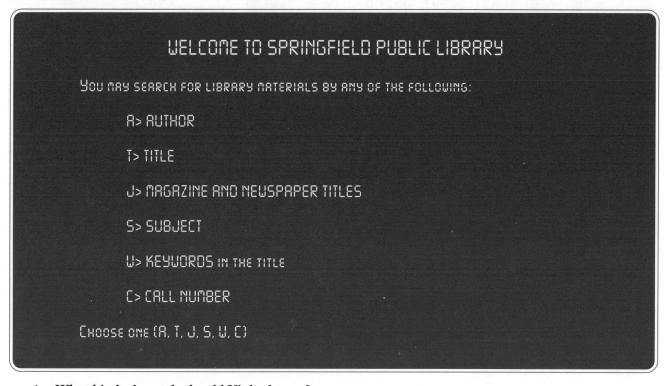

WELCOME TO SPRINGFIELD PUBLIC LIBRARY

You may search for library materials by any of the following:

A> AUTHOR

T> TITLE

J> MAGAZINE AND NEWSPAPER TITLES

S> SUBJECT

W> KEYWORDS in the title

C> CALL NUMBER

Choose one (A, T, J, S, W, C)

1. What kind of search should Vinh choose?
2. What key should he hit?
3. What are some keywords that Vinh could use?

```
You searched for the KEYWORD: WRITE RESUME

4 ENTRIES FOUND                                  LOCATION      CALL NO
1. How to write a job-getting resume (by)        NONFICTION    651.7    M   1991
2. How to write a winning resume/Deborah         OVERSIZE      650.14   B   1989
3. How to write a winning resume/Deborah         NONFICTION    650.14   B   1989
4. The resume solution: How to write (an         NONFICTION    650.14   H   1995

Please type the NUMBER of the item you want to see, OR
N> NEW search
A> ANOTHER search by KEYWORD
D> DISPLAY title or author
X> Sort by year
```

Answer these questions.

1. What keywords did Vinh type in?
2. How many books did he find?
3. Are all the books located in the same section of the library?
4. What key should Vinh hit to choose *How to Write a Winning Resume*?

```
AUTHOR:        BLOCK, Deborah Perlmutter
TITLE:         How to Write a Winning Resume
EDITION:       New Edition
PUBL & DATE:   Lincolnwood, Illinois: VGM Career Horizons, 1989
DESCRIPT       117 p. ill., 28 cm
SUBJECT        Resumes (Employment)

LOCATION            CALL NO.                STATUS
1>NONFICTION        650.14 B 1990           DUE 8-29-99
2>OVERSIZED         650.14 B 1990           DUE 9-10-99

Key NUMBER to see more information OR
Q> QUIT                      T> MORE TIME
N> NEW Search                A> ANOTHER search by KEYWORD
Choose one (1-2, Q, N, T, A.)
```

Answer these questions.

1. How many copies of the book does the library own?
2. What is the status of the book?
3. It is August 18 today. Is the book available?
4. Vinh wants to leave the computer now and go talk to the librarian. What key should he hit?

B. Match the meanings on the right with the abbreviations on the left. Write the letter of the meaning in the blank next to the abbreviation it matches.

Example: *publ.* is an abbreviation for *publisher,* so it is a match.

Abbreviations

_____ call no.

_____ publ.

_____ ©

_____ descrip.

_____ p.

_____ ill.

_____ cm

Meanings

a. publisher

b. description (notes about what the book looks like)

c. page

d. call number (for finding the book on the shelves)

e. illustrated (means the book has pictures or drawings)

f. centimeters

g. copyright (means that only the publisher can copy or print the book)

C. Read the situations below and decide what Vinh needs to do. Then look for the keyboard command that means the same thing and write the letter of the key he needs to choose for each situation. Then indicate on which screen the command can be found, screen 1 or screen 2. The first situation is done for you.

Commands on Screen 1:

You may search by:

A> AUTHOR

T> TITLE

J> MAGAZINE AND NEWSPAPER TITLE

S> SUBJECT

W> KEYWORDS

C> CALL NUMBER

Commands on Screen 2:

Key number to see more information, or:

N> NEW SEARCH

A> ANOTHER SEARCH BY KEYWORD

D> DISPLAY TITLE OR AUTHOR

X> SORT BY YEAR

Q> QUIT

T> MORE TIME

Situation	Keyboard Command	Screen
1. Vinh wants to look for the subject *job hunting.*	S	1
2. He wants to look for a book called *Better Jobs Today.*		
3. He knows the call number, but he doesn't know the location.		
4. He wants to search for books published after 1995.		
5. He wants more information on *The Resume Handbook.*		
6. He needs more time to copy the information from the screen.		

What Should I Do Now?

A. Listen to the conversation and check true (T) or false (F).

		T	F
1.	The book that Vinh wants will be available at a later date.	_____	_____
2.	Vinh cannot put a hold on any books at the library.	_____	_____
3.	He can go back and search for an available book.	_____	_____
4.	He cannot find books by looking for them on the shelves.	_____	_____

B. Answer the following questions.

1. When will the book be available?
2. The librarian makes three suggestions. What are they?

C. Listen to the conversation between Vinh and the librarian and fill in the blanks.

1. Vinh: What should I do now?

   Librarian: _____ a number of things. You _____ put a hold on the book. That way we won't lend it out again, and we'll call you as soon as the book comes back.

2. Vinh: Can I call in to see if the book is available?

   Librarian: Actually, we'll call you. But _____ take this brochure with you. It'll give you instructions on how to access our online catalog from home.

3. Vinh: Is there anything else I can do while I'm here?

   Librarian: _____ looking for similar books on the same shelf? _____ write down the call number for that book, and then look around the same area for another book. The call numbers are arranged on the shelves by subject.

D. Talk to another student. The first person asks about different problems, and the second person makes suggestions. Try to use many different forms of suggestions. Then switch roles.

Examples:	**STUDENT A:**	**STUDENT B:**
	How can I . . .?	Why don't you . . .
		Maybe you could . . .
		You might want to . . .
		Have you tried . . .?
		I suggest that you . . .

Searching for Specific Information

Vinh went back to the computer and searched under the subject *employment*. This search led him to a book called *Resumes That Will Get You The Job You Want*, by Andrea Kay. Since that book was not available, Vinh made another search under the same call number, 650.14K. He wanted to see if there were any more books on the same shelf, as the librarian suggested. He found a number of books, which are listed on the chart below.

Read each question below. Then look for the answers on the chart that follows. Look only for the specific information you need.

Questions	Answers
1. Who is the author of *Electronic Job Search Revolution*?	_____
2. Who wrote *Sure-Hire Resumes*?	_____
3. How many books on this list were written by Arnold B. Kanter?	_____
4. Which book contains the latest information?	_____
5. Which book by Robbie Kaplan was written first?	_____
6. How many entries can you find under call number 650.14 K?	_____
7. How many books are available now?	_____
8. How many books will become available on September 11?	_____

	Author	Title	Pub.	Call No.	Status
1.	Kanter, Arnold B.	The Essential Book Of Interviewing	1995	650.14 Kanter	9/11/99
2.	Kaplan, Robbie Miller	101 Resumes For Sure-Hire Results	1994	650.14 Kaplan	9/7/99
3.	Kaplan, Robbie Miller	Sure-Hire Cover Letters	1994	650.14 Kaplan	8/25/99
4.	Kaplan, Robbie Miller	Sure-Hire Resumes	1990	650.14 Kaplan	9/8/99
5.	Kay, Andrea	Interview Strategies That Will Get you The Job You Want	1996	650.14 Kay	Available
6.	Kay, Andrea	Resumes That Will Get You The Job You Want	1997	650.14 Kennedy	9/11/99
7.	Kennedy, Joyce Lain	Electronic Job Search Revolution	1994	650.14 Kennedy	Available
8.	Kennedy, Joyce Lain Lain	Electronic Resume Revolution	1994	650.14 Kennedy	Available

The Online Catalog, Partner 1

A. Work with a partner. The first person looks at the online catalog entry on this page. The second person looks at Page 96. Ask your partner for the information missing on your page. Write in the missing information. Your partner will ask you about information missing on Page 96. Look at your information to answer.

Example:

Partner 1:	What is the author's first name?
Partner 2:	Andrea. What is her last name?
Partner 1:	Kay.

AUTHOR   Kay, _____

TITLE   Resumes that will get you the job you want

EDITION   _____ ed.

PUBL&DATE   Cincinnati, _____: Betterway Books, _____.

DESCRIPT   166 p. : ill. ; _____ cm.

SUBJECT   _____ (Employment)

LOCATION	CALL NO.	STATUS
_____	650.14 Kay	DUE 07-_____-_____ (this year)

Key NUMBER to see more information, OR

    A > _____ Search by TITLE

____ > Browse nearby _____   Z > Show items nearby on _____

    N > NEW _____   ____ > ADDITIONAL options

Choose one (1, M, R, N, A, Z, S, P, +)

B. Now, together with your partner, compare answers. Did you both write in the missing information correctly?

Focus on Grammar

A. Put the words below in the correct order and write the sentences.

1. found / you / Have / job / yet / a ?

_____

2. for / I've / three / applied / already / jobs .

_____

3. resume / prepared / Have / ever / a / you ?

_____

4. computer / on / Have / searching / the / tried / you ?

_____

5. has / gone / library / She / to / public / never / a .

_____

6. son / a lot of / at / found / information / library / has / the / My .

_____

B. Study the chart. Then fill in the blanks with the correct forms of the verbs given.

Simple past forms: _did, didn't, went, checked,_ and so on	➔	Specific past times: _last week, a few minutes ago,_ and so on
Present perfect forms: _have/has, haven't, tried, written,_ and so on	➔	Nonspecific past times: _already, before, ever, yet,_ and so on

1. I _____ _____ at a lot of part-time jobs since I came here.
   (work)

2. There was nothing on the bulletin board when _____.

3. I _____ already _____ the classified ads.
   (try)

4. When I went to the library, I _____ how to look for a book.

5. Vinh _____ never _____ a resume before.

Focus on Communication 4

The Online Catalog, Partner 2

A. Work with a partner. The first person looks at the online catalog entry on Page 94. The second person looks at this page. Ask your partner for the information missing on this page. Your partner will ask you about information missing on Page 94. Look at your information to answer.

Example:

Partner 2: What is the author's last name?

Partner 1: Kay. What is her first name?

Partner 2: Andrea.

AUTHOR   _____, Andrea

TITLE   _____ that will get you the job you want

EDITION   1st ed.

PUBL&DATE   _____, Ohio: _____, 1997

DESCRIPT   _____ p. : ill. ; 28 cm

SUBJECT   Resumes (_____)

LOCATION   CALL NO.   STATUS

NONFICTION   _____   DUE _____ -25- _____
                                                                               (this year)

Key NUMBER to see more information, OR

    A > ANOTHER Search by _____

    R> _____ nearby entries   _____ > Show items nearby on shelf

    N > NEW search   +> _____ options

Choose one (1,   M,   R,   N,   A,   Z,   S,   P,   +   )

B. Now, together with your partner, compare answers. Did you both write in the missing information correctly?

Word Search

Find as many words as possible in the grid below. Hint: They are all related to the library. When you find a word, spelled across or down, circle it. There are 28 words all together.

```
P  A  I (S  U  B  J  E  C  T) T  H  O  C  R  E  L  M  V  L
L  E  N  O  R  S  U  D  U  E  D  A  T  E  A  S  T  E  R  O
I  C  A  T  R  E  L  I  B  R  A  R  I  A  N  O  N  L  U  G
A  N  B  I  E  L  R  T  S  W  A  U  T  H  O  R  S  D  A  T
S  O  R  P  L  E  A  I  K  T  L  O  F  B  A  T  E  N  I  O
E  N  T  R  Y  B  L  O  C  A  T  I  O  N  K  O  S  C  I  P
B  F  I  C  T  I  O  N  L  A  C  T  I  T  L  E  T  O  B  A
T  I  T  P  O  R  T  R  A  W  H  E  H  O  K  E  A  F  R  E
A  C  H  E  C  K  O  U  T  D  E  S  C  R  I  P  T  I  O  N
O  T  H  L  A  S  E  L  E  F  P  U  B  K  E  W  U  N  W  D
S  I  C  S  L  E  N  D  R  O  D  A  S  O  S  Y  S  E  S  T
E  O  N  L  I  B  R  A  R  Y  C  H  A  E  P  R  E  E  C
B  N  O  I  N  A  P  O  N  L  I  N  E  C  A  T  A  L  O  G
F  I  C  P  U  B  L  I  S  H  E  R  L  A  R  P  K  O  T  S
S  A  K  R  M  E  N  T  Y  C  O  O  F  B  C  A  L  A  N  D
B  R  A  D  B  O  N  X  A  M  P  V  T  O  H  I  F  N  I  P
G  O  P  K  E  Y  W  O  R  D  N  J  B  C  X  O  R  T  M  I
L  E  C  A  R  D  C  A  T  A  L  O  G  R  A  P  M  E  L  F
```

Role Play

1. Imagine that you are a librarian. The other students in your group are visitors. Answer all the visitors' questions about different ways to search for a book, due dates, and library policies. Answer their questions about what they can do if the book they want is not available. Then switch roles and have another person be the librarian.
2. Imagine that your friend is looking for a job. This friend has never worked in this country before and has never looked for a job here, either. Explain your friend's situation to your group and ask everyone in the group to give advice and make suggestions for your friend.

Word Search Answers

Did you find all these words in the Word Search on Page 97?

subject	due date	librarian	card catalog	edition	title	search
author	nonfiction	call number	location	fiction	browse	fine
description	online catalog	shelf	checkout	lend	loan	library
publisher	keyword	status	entry	sort	late	slip

Coping Skills

In order to learn how to find a book at the library, Vinh took a variety of steps.

A. Put a ✓ next to the steps he took to achieve his goal. Give examples.

( )   described a problem

( )   asked for advice

( )   asked for help

( )   asked about options

( )   considered alternatives

( )   read on-screen instructions

( )   located specific information on the computer

( )   followed instructions on the computer

( )   followed spoken instructions

B. With a group, discuss the coping skills above.

What is the most important skill you learned in this unit? Which ones will you use in the future? In what situations will you use them?

Community Assignment

Go to your local library. Do a search by subject and find a book on employment, resumes, or any other subject you are interested in. If your library does not have an automated system, use the card catalog. Look up a book and write down the title, author, and call number from the information in the catalog. Then bring the information to class. In a group, look at the titles and decide which book each person wants to check out.

How May I Direct Your Call?

Have you ever worked in an office? What did you do there? How are offices in this country different from offices in your country?

The Front Office

This is Barbara. She works in the front office here at Dominie Press. She is a receptionist and general office clerk. She takes orders for books and educational materials and sees that they are filled quickly and accurately. She's very organized and courteous. She knows how everything is done around here, from the time we receive an order until the books are finally packed and shipped to customers all over the United States and Canada. We couldn't get along without her!

Read the story and answer these questions.

1. What is Barbara's position at Dominie Press?
2. Who do you think is introducing her?
3. Is she a good worker?
4. Define the following: *accurately; organized; courteous;* and *get along without her.*
5. Define the following: *pack* and *ship.*

Look and Listen 👁 👁 👂

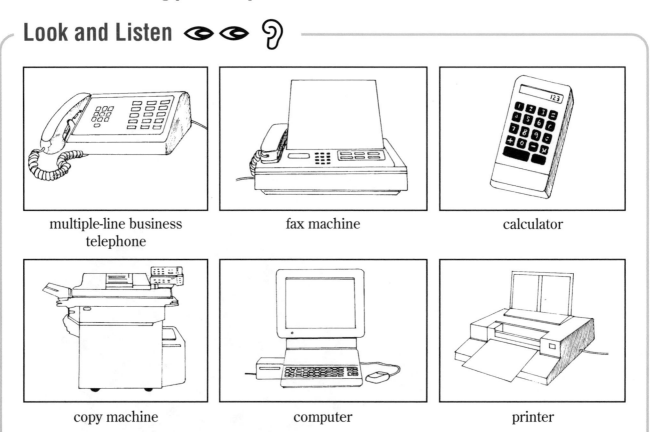

| multiple-line business telephone | fax machine | calculator |
| copy machine | computer | printer |

Listen to Barbara describe her day yesterday. Then write the name of the piece of office equipment that matches each description. Listen as many times as necessary.

1. The _____ was ringing all day.

2. The _____ needed service.

3. The _____ was out of paper.

4. The _____ was down.

5. The _____ was missing from her desk.

6. The _____ was jammed.

Focus on Listening 1

Take-Our-Daughters-to-Work Day

Today is Take-Our-Daughters-to-Work Day. This is a special day every year when parents bring their daughters to work with them. The idea is to let young girls see what their mothers and fathers do at work and to show them different kinds of work that they might choose in the future. Barbara has brought her granddaughter Lisa to work today, and she's showing her around the office.

A. Listen to the conversation between Barbara and Lisa and check true (T) or false (F).

	T	F
1. Lisa likes the office.	____	____
2. She thinks answering the phone looks easy.	____	____
3. All the office equipment is working well today.	____	____
4. Lisa is ready for a lunch break.	____	____

B. Answer the following questions.

1. What did Lisa do to help?

2. What did Lisa think of Barbara's job?

3. Would you like to have a job like Barbara's?

C. Listen to Barbara giving Lisa instructions on how to use the copy machine. Then listen again and write down the steps for making a copy on the lines below. Listen as many times as necessary.

1. Lift _____

2. Put _____

3. Select _____

4. Press _____

5. Take _____

D. Compare your instructions with another student's instructions and check your spelling. If there is anything you're not sure about, check with your teacher.

Vocabulary

Barbara used words with the following meanings when she was answering the phones. Match the words on the right with the meanings on the left. Write the letter of the word in the blank next to the meaning it matches.

Meanings	Words
_____ answer the telephone in an office	a. transfer
_____ wait, stay on the phone line	b. call back
_____ switch a phone call to another line	c. take a message
_____ tell the receptionist your name and phone number	d. cover the phones
_____ write down the caller's name, phone number, and any other important information	e. voice mail
_____ call the person later, return the phone call	f. hold, hold on
_____ an automated system for answering the phone and taking messages	g. leave a message

Focus on Communication 1

I'm Sorry, She's Away from Her Desk

Receptionists use expressions like "Please hold" or "I'll transfer you" all the time, and it's not too difficult to know what they mean. However, there are some other polite expressions that office workers use on the phone, like "stepped out," "gone for the day," and "with a customer." These expressions have special meanings that are not obvious. They are used to give the caller more information about someone who's not able to answer the phone, such as how long the person will be busy, if the person has already gone home, or when the person will be able to call back. Do you know these mystery expressions?

A. Work with a partner. Read the polite expressions on the left and try to guess what they mean. Write the letter of the meaning on the right in the blank next to the polite expression it matches.

Polite Expressions

_____ I'm afraid she's gone for the day.
Can I transfer you to her voice mail?

_____ I'm sorry. He's stepped out.
Can I take a message for him?

_____ He's with a customer right now.
Would you like to leave a message?

_____ She's on the other line right now.
Shall I ask her to call you back?

_____ I'm sorry. She's away from her desk.
Can I have her call you back?

_____ He isn't in right now.
Can I have him call you after lunch?

_____ I'm sorry. He's on the other line.
Would you like to leave a message
on his voice mail?

_____ She's in a meeting right now.
Can I have her call you later today?

Real Meanings

a. He's somewhere in the office, but I'm not
    sure where.

b. He's eating, and he'll be working again in
    thirty minutes to one hour.

c. She's busy talking to her co-workers about
    something important.

d. She's already gone home. It's too late to
    reach her today.

e. He's busy working with somebody else right
    now.

f. She left the office to go to the ladies room,
    but she'll be back soon.

g. She's talking on the phone to someone else
    right now.

h. He's talking on the phone right now. He'll call
    you back after he's finished.

How to Transfer a Call

When Lisa was listening to Barbara on the phones, it seemed like a really difficult job to use the phone system to transfer calls to different offices. But later when Lisa was sitting at Barbara's desk, she noticed a neatly typed note near the telephone.

A. Read Barbara's note and answer the questions below.

To transfer a call:

Press TRANSFER.

Dial the extension.

Wait for the person to answer.

Tell the person who's on the line.

Hang up.

To forward calls to the voice mail system:

Dial *72.

Wait for the dial tone.

Dial the telephone number for the voice mail line.

Wait for the system to answer.

Hang up.

To bring the incoming calls back from the voice mail system, dial *73.

1. Is it difficult to transfer a call?
2. What do you need to ask the caller before you transfer the call?
3. What happens when you hang up?
4. When do you think Barbara uses the "forward calls" procedure?
5. When do you think she needs to bring the calls back?

B. In the conversation below, Barbara is transferring a call to Mr. Perez. Fill in the blanks.

Barbara:	Good morning. Dominie Press.
Caller:	_____. Is Mr. Perez _____, please?
Barbara:	Yes, he is. _____ calling, please?
Caller:	This _____ Ray Price.
Barbara:	Hold _____, please. (She dials Mr. Perez's_____.)
Mr. Perez:	Yes?
Barbara:	Ray Price is _____ the line for you. (She hangs _____.)
Mr. Perez:	Ray! How are you?

Focus on Communication 2

Could You Help Me Out with Something?

Words	Meanings
order processing	filling out order forms and putting information in the computer
accounting	keeping records of how much money comes in and goes out
department	one office or section of a company or organization
purchase order	an order that will be paid for after it is delivered
mixed up	not organized; not in order

When Barbara and Lisa came back from lunch, Lisa noticed something different on Barbara's desk.

A. Listen to their conversation and answer these questions.

1. What was different? What did it mean?
2. What did Barbara do?
3. What did she tell Lisa to do?
4. What was the message about?
5. What did the order processing clerk ask Barbara to do?
6. What was wrong with the order?

B. Listen to Barbara calling John to give him the message. Listen as many times as necessary to fill in the blanks.

Hello, John? This is Barbara. I just called Mila at Edwards Community Adult School, and they

_____ you _____ _____ the order. She _____ me _____ _____ you

know. She will _____ the school district accounting department _____ _____ a new

purchase order. She _____ _____ _____ you it's important to make sure you cancel

the first order, because if there are two different purchase orders, everything gets all mixed up.

That's right, cancel purchase order 1609734. Okay, you're welcome. Bye.

C. Put the words in the correct order and write the sentences about the phone calls on the lines below.

1. Lisa / Barbara / a / hand / told / message pad / to / her .

   _____

2. help / asked / Barbara / to / John / out / something / with / him .

   _____

3. wanted / call / to / He / customer / the / Barbara .

   _____

4. department / the / cancel / to / order / customer / wanted / order / The / the / processing .

   _____

5. message/ John / give / Mila / asked / to / Barbara / the .

   _____

D. Listen to John's message again. Listen to the way John asked Barbara to do him a favor and fill in the blanks.

1. Hi, Barbara. It's John. Listen, could you _____ _____ a favor? I need you _____
   _____ me out with something.

2. So, anyway, _____ you _____ calling the contact person there for me? I just don't
   have time to take care of it today.

E. Work with a partner. The first person asks a favor and explains what favor is needed. The second person responds. Then switch roles.

Example: **Student A:**   _____, could you do me a favor?
                                       (Student B's name)

                  Student B: Sure, what is it?
                  Student A: I need you to help me out with something. Would you mind
                                   _____ for me?

                  Student B: Sure, no problem. I can take care of that for you.

                  Student A: Oh, thanks very much. I appreciate it.

Focus on Listening 2

Thank You for Calling

Words	Meanings
employee	someone who works at this place
extension	a code for calling a particular person in an office
digit	a single number; for example, a telephone number has seven digits
recording	a message that is recorded on tape
reach	connect with someone by telephone
party	person; for example, the person you are calling
counseling office	a school office where students get information about classes
administrative office	the office of a school principal or company president
plant manager	a person who takes care of a building or buildings
assistance	help

When Barbara called Edwards Community Adult School, she heard a voice mail recording.
She didn't know Mila's extension; only that she worked in the school bookstore.
She had to listen to a long voice mail recording and wait for the instruction that would tell her how to reach Mila.

A. Listen to the school's voice mail message. In the blanks, write the number you need to press to reach each of the following people or offices.

1. To reach the school principal, press _____.
2. To get information about English classes, press _____.
3. To speak to someone in the plant manager's office, press _____.
4. To reach the operator, press _____.
5. To speak to Mila in the bookstore, press _____.
6. To avoid the voice mail system, press _____.

B. Listen to the school's voice mail message again and fill in the blanks.

Thank you for _____ Edwards Community Adult School.

If you are an employee calling in an absence, please press _____.

Our office hours are Monday through Thursday from _____ A.M. to 9:30 P.M., and Friday from

8:00 A.M. to _____ P.M.

If you know your party's three-digit _____ and wish to leave a message, you may

_____ it at any time.

For information regarding _____ classes, please press 2.

For the counseling _____, please press 3.

For the administrative offices, please press _____.

For the _____, please press 5.

To _____ to the plant manager, please press 6.

All other callers, please dial the _____ for assistance or remain on the _____.

C. Work with a partner. Read the voice mail message above. Then write a similar voice mail message for your school or business. Create a message for an imaginary business if you wish. Write the script for your recording below.

_____

_____

_____

_____

_____

_____

_____

Focus on Information

Learning a New Procedure

Later the same afternoon, Barbara took Lisa to the order processing department at Dominie Press. She wanted to show her how the data entry clerks enter information, or data, on the computer. Barbara and Lisa are talking to John, who works in the order processing department.

A. Read the conversation between John and Lisa, and then read the computer screen below.

John: Hi, Lisa. Nice to meet you. Do you want me to show you what I'm doing here?

Lisa: Yes, please.

John: I'm entering this order in the computer. We got the purchase order from a school in Michigan, and now I'm processing the order so that we can ship it out as soon as possible.

Lisa: What information are you entering in the computer?

John: This is a regular customer, so they have a customer number. I just enter the customer number where it says "Bill To" right here, and then the computer automatically displays the name and address of the school.

Lisa: Oh, I see. Does "Bill To" mean who will pay for the order?

John: Yes. Then if the information is correct, I hit enter, and the next window opens.

Lisa: Oh, wow!

```
BILL TO:    40517                           SHIP TO:  40517



          ROMULUS COMMUNITY SCHOOLS
          36540 GRANT ROAD
          ROMULUS  MI 48174
```

B. Read the continued conversation between John and Lisa and the computer screens below.

John: This is where I enter the shipping information, the address where we will send the books. I type in the address, and then there's a question at the bottom of the screen. It says, "All OK?" See that? So I answer yes by typing Y and then . . .

Lisa: Oh, look!

John: This is a new window. Whenever the computer gives you a new space on the screen, that's called a window. This is the order entry window. I just start typing in the order here. First I need the item number for each book. That's the Dominie Press item number from our catalog. Then I type in the quantity. That means how many books they want to order. Then usually the description and price come up automatically, and I go on to the next item.

Lisa: How do you know where to type the data?

John: Well, each time you hit ENTER, the cursor moves to the next place you need to enter something.

Lisa: Did you say the cursor?

John: Yes. The cursor shows you where you are on the screen. You see that little line in the box? That's the cursor. If I begin typing right now, that's where the information will be entered. If I want to move to a different place, I use the arrow keys on the keyboard. I can move up, down, right, or left on the screen.

Lisa: Oh, I see. Was it hard to learn how to enter the items?

John: No, not really.

BILL TO:	ROMULUS COMMUNITY SCHOOLS			QUAN. FOR ORDER	=	12

Book	Quan.	Description	Price	Dis%	Net $
ELC459	4	MAGNETIC LETTERS LC 45 PCS IN JAR	$ 5.75		23.00
ELV309	2	MAGNETIC LOWER-CASE VOWELS (30 P(CS)	$ 3.95		7.90
MAGBRD	3	MAGNETIC SPELL BOARD	$ 3.00		15.00
DACH	1	DESKTOP ALPHABET CHART — 10 PACK	$ 5.00		5.00
SHP		***PLEASE NOTE: SHIPPING IS A MIN. $#4.50 FOR ORDERS UNDER $45.00, AND 10% THEREAFTER.			

John: Now watch. When I'm finished with the order, I hit ENTER two times. Now there's another question at the bottom of the screen.

Lisa: It says, "Any error in book item entry?"

John: Right. So now I check the items and make sure they're the same as the purchase order, and then I hit the N key to answer no. There are no errors—no mistakes in the list of items.

Lisa: Look what happened! A new window opened. What is that?

John: That's more shipping information. It looks fine, so now I press ENTER again. And here's the last screen. This is the Final Closing screen. Now you can see the invoice number the computer gave to this order. That's the number we use if we need to check the order later. Also, the accounting department has to send an invoice with this number on it to the school.

Lisa: The bill is called an invoice?

John: Yes. Now I'm ready to print out this order. Then I'll give one copy to the accounting department so that they can send the invoice, and I'll give one copy to the shipping department so that they can send the books out to the school.

```
Final Closing . . . . . .
                . . . . . . . . . . . . . . . . . . . . .
                . Invoice Number: 159843 .
                . . . . . . . . . . . . . . . . . . . . .
        Value of this invoice:

                Net $ for all items    -    $        50.90
                Shipping charges       -    $         5.09
                Sales Taxes (if any)   -    $          .00
                                            ——————————
                                            $        55.99
```

C. Match the words on the right with the meanings on the left. Write the letter of the word in the blank next to the meaning it matches.

Meanings	Words
_____ type in	a. information
_____ Bill	b. space on the screen
_____ hit ENTER	c. up, down, right, left keys
_____ window	d. enter
_____ data	e. take care of an order
_____ send	f. this customer will pay for the order
_____ process an order	g. press ENTER
_____ arrow keys	h. error
_____ mistake	i. ship
_____ Bill to: Main St. School	j. invoice

D. Look at the data entry windows below. Imagine that you are an employee at Dominie Press, and John has just taught you how to enter the orders on the computer. For each window, write notes to help you remember what to do. Then compare notes with a partner.

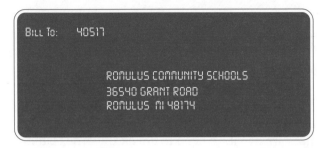

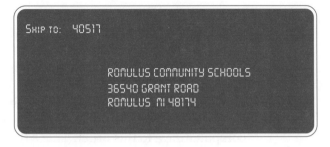

_____

_____

_____

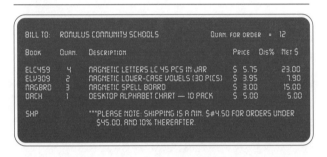

_____

_____

_____

Focus on Grammar

A. Work with a partner. The first person thinks of a question with *ask, tell, want,* or *need,* as in the examples below. The second person answers. Then switch roles.

Example:	**Student A:**	What does your brother want you to do?
	Student B:	He wants me to go to his house this Sunday.
	Student A:	What did your friend ask you to do?
	Student B:	She asked me to give her a ride to work.

B. Choose three of your most interesting sentences from exercise A and write them on the lines below. Then ask your partner or your teacher to check your sentences.

1. _____

2. _____

3. _____

Focus on Reading

Observing People at Work

Words	Meanings
observe	watch; pay attention
co-workers	people who work together
"get to"	have the opportunity to do something
interact	talk with each other
judged	recommended; described
interpersonal skills	ability to interact with people
cooperative	a person who wants to work with others
team player	a person who can work well with a group
respect	understand; think well of other people
"get along with"	be friendly; communicate well with others
"people skills"	communication and interpersonal skills

When I brought Lisa to work with me, she got to see a lot of things in the office. My co-workers and I showed her how to use the phones and all the office equipment. John even showed her how to enter an order on the computer. She knows how the company is organized, who does what, and how we serve our customers. I'm sure it was interesting to her.

But I was thinking about this. I think that the most important thing that Lisa saw was how we work together—how we interact with each other and help each other. That is one of the most important things to learn about working. The way you work with people is how you will be judged as an employee. I think that's true no matter what kind of work you do. It's not just when you're working with customers; it's when you're talking with co-workers, the supervisor, the boss, anybody. Everyone likes to work with a pleasant, cooperative person. You have to have good interpersonal skills.

You have to be a team player, too. This means that you respect your co-workers; you listen to them and show that you respect their ideas; and you show them that you want to work together well. I think getting along with people is the most important thing on the job. I hope Lisa observed some good people skills when she visited us at Dominie Press today.

A. Read the story and check true (T) or false (F).

		T	F
1.	Lisa saw many different people doing their jobs.	_____	_____
2.	Barbara talked to her about how to work with people.	_____	_____
3.	Good interpersonal skills are important only for working with customers.	_____	_____
4.	Co-workers should work together well.	_____	_____

B. Work with a partner. Discuss the story on Page 113. Talk about what you believe are the most important skills to remember when you are working with others.

Role Play

Imagine that you are applying for a job. You want to explain that you are a good worker because you know how to work well with others. Write a few notes first, and then practice telling your group how you would answer the question, *Why should we hire you?* in an interview. Ask your group to judge you on how well you spoke and how well you presented yourself in your practice interview.

Coping Skills

When Lisa visited the Dominie Press office, Barbara and John showed her how they do their jobs. What did she observe them doing at work?

Put a ✓ next to the things she observed. Give examples.

( )   answered the phone

( )   transferred a call

( )   left a voice mail message

( )   retrieved a voice mail message

( )   followed instructions on a voice mail recording

( )   entered data on a computer

( )   gave someone instructions

( )   asked for a favor

( )   showed someone how to do something

Community Assignment

Call a store, school, or office in your community when they are closed. If you hear a voice mail recording, write it down. (You may need to call several times to write down the message.) If there are several choices of numbers to press, write down what number to press for different choices. Bring your notes to class and share them with your group. Then, together with your group, make a list of all the places in your community that have voice mail recordings on the phone.

Packing and Shipping

What kind of place is this? What kinds of jobs are there in this kind of business?
What skills do you need to work in a place like this?

Plumbing Parts Distribution Center

This warehouse is a large distribution center for a company that sells parts for plumbing; that is, parts for installing and repairing sinks, showers, water pipes, garden sprinklers, and so forth. There are thousands of different parts here. Stores and plumbing supply companies order the parts they need, and this company ships large orders to them. Some of the jobs here require physical strength, the ability to drive a forklift, office skills, or management skills. Almost all of the jobs here require some knowledge of computers.

Answer these questions about the company in the story.

1. What does this company sell?

2. What kind of building is this?

3. What skills do workers need for some of the jobs in a warehouse?

4. Do you have any of those skills?

5. Why is knowledge of computers important in a company like this?

Focus on Reading 1

Warehouse Jobs

There are many different jobs in a large distribution center. There are truck drivers and forklift drivers. There are cleaning people and maintenance people. There are managers, office workers, accounting clerks, and payroll clerks. But the largest number of workers are the order pullers and packers.

Every warehouse has some kind of product in storage. The company's job is to keep the product in an organized place, and then distribute the product when there is an order to ship it someplace else. The order is called a transfer order, and the distribution company will transfer, or ship, the product by truck or other transportation.

The job of the order puller, or order selector, is to *pull* the items off the shelves to fill an order. The order puller uses a list of the items and quantity needed and the location of the items in the warehouse. The locations are organized by areas, rows, and bins. *Bin* means a space on a shelf or, for smaller items, a space in a box or drawer. The items, areas, and bins all have codes–combinations of letters and numbers which identify them. The transfer order shows all the codes for the items and their locations. The job of the order puller is to find all the items, remove them from the bins, and take them to the packing and shipping area. An order puller or order selector can also be called an order picker or warehouseperson.

The job of a packer is to put an order together in a larger box, or carton. The transfer order must be checked again and any missing items marked. The items that are ready to be shipped are packed on a *packing line*. The packing line is a belt that moves the boxes from a packing area to a scale for weighing the boxes and preparing a shipping label. After the shipment (one or more boxes that are prepared for shipping) has been weighed, the packer moves it down the line again and puts it on a pallet. The pallet is a square of wood that can be picked up by a forklift. The packer puts cartons on the pallet, and then the forklift driver moves the whole pallet onto a truck. If the items and cartons are smaller, the packer may prepare them for shipping and then put them directly in a mail bin so that they can be sent out by regular mail or picked up for delivery by a shipping company.

A. Read about warehouse jobs again and find the meanings of the following words.

storage	bin	carton
transfer	code	packing line
transfer order	order picker	shipment
order puller	packer	pallet

Circle each word in the story on the previous page, and underline the part of the story that tells you what it means.

Look and Listen 1 👁️ 👁️ 👂

A. Look at the pictures and listen to the tape.

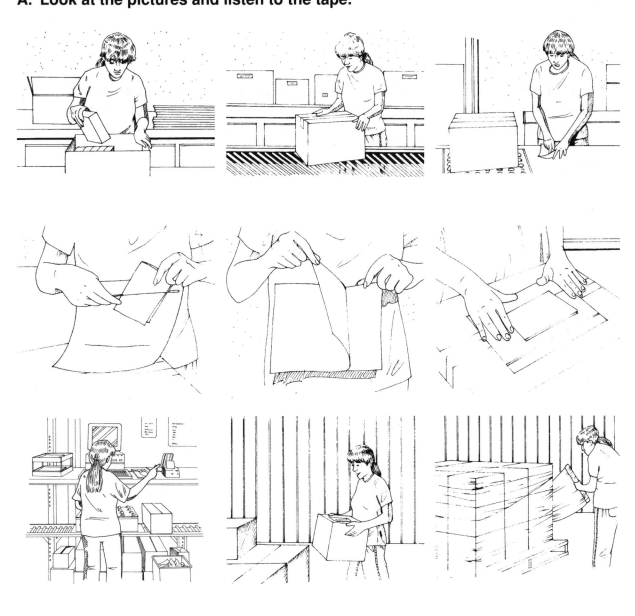

B. Read the following sentences and listen to the tape again.

1. Pack the order in a carton.

2. Roll the carton down the line.

3. Fold the order ticket with the address facing out.

4. Slip it into a plastic sleeve.

5. Peel the backing off the plastic sleeve.

6. Stick it on the carton.

7. Tear off a shipping label and stick it on the carton.

8. Load the carton onto the pallet.

9. Wrap the loaded pallet with plastic.

C. Act out the sentences as another student says the steps. Then switch roles. Look at the pictures on the previous page. Act out the sentences again and switch roles again.

D. Write sentences for all the actions* you just did.

1. *I packed the order in a carton.*

2. *I rolled the carton down the line.*

3. _____

4. _____

5. _____

6. _____

7. _____

8. _____

9. _____

*Note: All of the past-tense verbs end in *-ed* except *stick/stuck/stuck/* and *tear/tore/torn.*

The Packing Line

Jose Morales has just been promoted to a better position at work. He was an order picker before, but he never used the computer. Now he is going to learn the packer's job and work on the computer on the packing line. Charles, his new supervisor, is going to teach him the job. Listen as Charles explains the packing line to Jose.

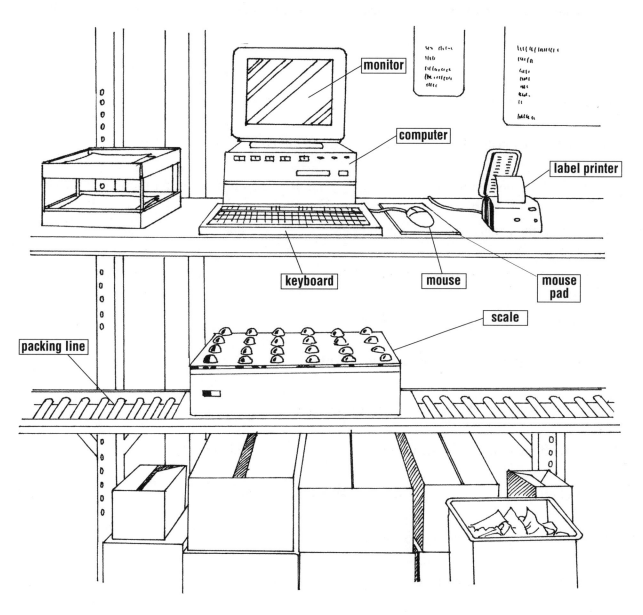

A. Listen to Charles teaching Jose how to work on the packing line. Then listen again and fill in the blanks.

1. First you need to check the order that was pulled. Count the _____ , and check the

   transfer order to make sure everything is here.

2. Make sure you have two pieces of paperwork: the transfer _____ and the delivery

   _____ .

3. Pack the order into a _____ .

4. Roll the carton down the _____ and put it on the scale.

5. Enter the order number, _____ number, and purchase order number in the computer.

6. Wait for the computer to print out the _____ label.

7. Meanwhile, fold the transfer order with the _____ facing out, like this.

8. Slip it into the _____ sleeve.

9. Stick it on the _____ of the carton.

10. When the printer stops, tear off the shipping _____ .

11. Peel off the top part and stick it on the _____ of the carton.

12. Peel off the _____ part and stick it on the delivery note.

13. Put the completed delivery note in this _____ .

14. Push the carton down to the _____ of the line.

B. Work with a partner. The first person gives all the instructions. The second person acts out the instructions without looking at the book. Then switch roles.

Vocabulary

check	make	roll
count	pack	slip
enter	peel off	stick
fold	print out	tear off
	put	wait

A. Fill in the blanks below with the correct verb.

1. First you need to _____ the order that was pulled. _____ the items and check the transfer order to _____ sure everything is here.

2. _____ sure you have two pieces of paperwork: the transfer order and the delivery note.

3. _____ the order into a carton.

4. _____ the carton down the line and _____ it on the scale.

5. _____ the order number, customer number, and purchase order number in the computer.

6. _____ for the computer to _____ _____ the shipping label.

7. Meanwhile, _____ the transfer order with the address facing out, like this.

8. _____ it into the plastic sleeve.

9. _____ it on the top of the carton.

10. When the printer stops, _____ _____ the shipping label.

11. _____ _____ the top part and _____ it on the side of the carton.

12. _____ _____ the bottom part and _____ it on the delivery note.

13. _____ the completed delivery note in this tray.

14. _____ the carton down to the end of the line.

B. Write sentences about the actions* Charles did. Choose six of the instructions above and write the sentences on a separate sheet of paper.

Examples:
1. Charles rolled the carton down the line.
2. He folded the transfer order with the address facing out.

*Note: All of the past-tense verbs end in *-ed* except *put/put/put* and *stick/stuck/stuck*.

Focus on Listening 1

Point and Click

Words	Meanings
command	an instruction that tells a computer what to do
mouse	a tool for using a computer by selecting commands on the screen
mouse pad	a soft pad on the desk that is needed for using a mouse
cursor	a small, blinking line (looks like |) on the screen that indicates where you can type next
pointer	another cursor (looks like →) that shows where you are moving on the screen when you use the mouse
point	move to a new place on the screen, using the mouse
click	press the button on the mouse
button	a square on the screen that you can click on to select a command
toolbar	a row of buttons on the screen that gives you a choice of commands

Charles is telling a story about how he felt when he first learned to use the computer at work.

A. Listen to Charles' story and check true (T) or false (F).

	T	F
1. Charles was excited from the beginning about learning the computer.	_____	_____
2. Charles is a big man.	_____	_____
3. At first he felt silly using the mouse.	_____	_____
4. Charles learned to *point and click* with the mouse.	_____	_____
5. Now he uses the computer often.	_____	_____
6. He still doesn't think the computer is very useful at work.	_____	_____

B. Look at the drawing of the computer screen on the next page and find the pointer, cursor, toolbar, and Ship Now button.

Real-Life Reading

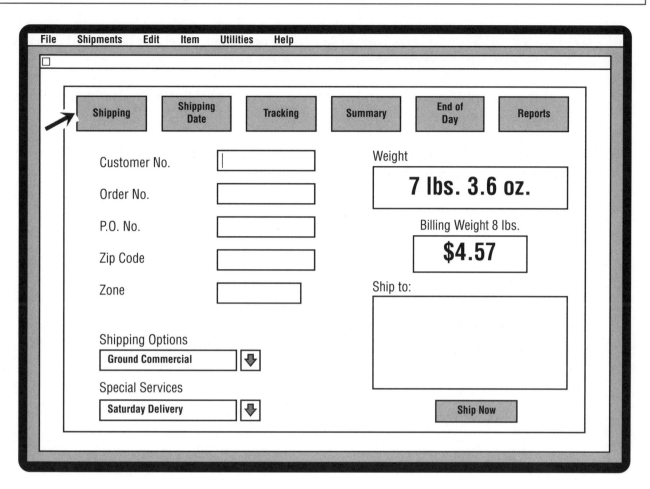

1. What commands do you see on the toolbar?

   _____

2. Which items have empty windows where you can type in information?

   _____

3. Which of the following shipping options has been selected?

   _____

Ground Residential
Ground Commercial
Air/International

4. Which of the following special services has been selected?

   _____

Next Day Express
Two-Day Air
Saturday Delivery

Look and Listen 2

When Jose was learning to work on the packing line, he didn't know anything about using a computer. Charles explained that he didn't know how to use it in the beginning, either, but he learned it on the job. He told Jose to watch and listen while he did a few shipments.

A. Look at the computer screen on the previous page and listen to the instructions that Charles gave Jose.

B. Listen to Charles' instructions again. Number the instructions below in the order that Charles told Jose to do them.

_____ a. Then you type in the customer number.

_____ b. Now you check all your numbers and make sure they're correct.

_____ c. Then you click on the Shipping button, and this screen comes up.

_____ d. Then go to the next box, click inside the box, and type in the P.O. number.

_____ e. When you see the insert cursor, you click right there.

_____ f. First you put the package on the scale.

_____ g. Then you click on the Ship Now button to process the shipment and print out the label. That's it!

_____ h. Then you put the pointer in the Customer No. box, near the left side.

_____ i. After that, you go to the next box. Click inside the box, and type in the order number.

Focus on Listening 2

What About These Boxes?

When Charles was showing Jose how to use the computer, Jose noticed that these two boxes were a little different.

Shipping Options

| Ground Commercial | ⬇ |

Special Services

| Saturday Delivery | ⬇ |

A. Listen as Charles explains how to use these two boxes, and fill in the blanks. Listen as many times as necessary.

Jose: What about these boxes? How _____ _____ _____ information in here?

Charles: Well, these both have menus. I'll show you. What you do is, _____ _____ on the arrow, but don't release the button on the mouse. Then a list, or menu, of different options will appear. _____ _____ the button down, and then _____ _____ the mouse up or down to select one of the options. _____ _____ _____ _____ type anything in.

Jose: Oh, I see.

B. Discuss the following questions with a partner or with your teacher.

1. Who is Charles talking to?
2. What does he mean when he says *you*? Jose? Someone? Anyone? Everyone?
3. Who might do the actions that Charles is describing?
4. Could you use the same expressions to talk about how any person would do the job?

C. Work with a partner. The first person gives instructions for using the shipping computer. The second person indicates understanding and asks questions if necessary. Look at the computer screen on Page 123 if you wish.

Example:	**Student A:**	First you put the box on the scale.
	Student B:	Okay.
	Student A:	Then you type in the customer number.
	Student B:	Uh-huh.
	Student A:	(Continue explaining the steps.)

Guessing Game

A. Computer words for the guessing game

click	mouse	mouse pad	point
pointer	cursor	button	command
toolbar	arrow	keyboard	printer
monitor	data	window	menu

B. How to prepare the game

- Form groups of three to five.
- Divide an 8 $\frac{1}{2}$-by-11-inch sheet of paper into sixteen sections and cut them apart.
- Write one of the above words on each section of paper to make cards.
- Mix up the cards.
- Place all the cards face down in one stack in the middle of the table.

C. How to play the game

- Players take turns.
- The first player picks up a card.
- Without saying the word on the card, the first player explains or describes the word.
- All others guess the word.
- When someone guesses the word correctly, the first player gives him or her the card.
- Each player picks up a card and explains or describes the word, and the others guess.
- When there are no cards left in the middle, all players count their cards.
- The player with the most cards is the winner.

Did You Click Inside the Box?

When Jose tried using the mouse for the first time, he had a few problems. Sometimes he missed a step, and then he couldn't figure out what was wrong. Charles was always able to help him. Usually he just asked Jose if he had done the previous step. For example, when Jose said he couldn't figure out why his typing wasn't in the right box, Charles asked, "Did you click inside the box?" When Jose couldn't find the box for the customer number, Charles asked, "Did you click on the Shipping button first?"

A. Read each problem below. Then write a question to help the person figure out the mistake and correct it.

1. Why isn't my typing in the right box?　　　　　(Click inside the box.)
   Did you click inside the box?

2. I can't find the box for the customer number.　　(Click on the Shipping button first.)
   Did you click on the Shipping button first?

3. Why didn't the mouse work correctly?　　　　　(Remember to point and then click.)

4. Why didn't the selection in this box change?　　(Move the pointer to the new selection.)

5. Why did the computer "beep" at me?　　　　　(Wait for the insert cursor.)

6. Why didn't the label print out?　　　　　　　(Check the address and zip code.)

B. Work with a partner. The first person asks about each step. The second person answers, "Yes, I did that" to several questions, or, "No, I didn't do that. Maybe that's what the problem is!" Then switch roles.

Example:	Student A:	Did you point to the box?
	Student B:	Yes, I did that.
	Student A:	Did you wait for the insert cursor?
	Student B:	No, I didn't do that. Maybe that's what the problem is!

Focus on Reading 2

Training and Improving Your Skills

Words	Meanings
skills	things that you are able to do, especially things that are useful at work
training	learning job skills
retraining	learning new ways to do your job or ways to do a new job
industry	type of business
development	progress, advancement, moving forward
opportunity	a chance to do something different
take advantage of an opportunity	benefit from an opportunity

In workplaces today, new technology has affected nearly every worker in every industry. Many industries have changed their procedures and equipment as new technology becomes available. Companies are always looking for a way to do the same work faster, better, and more easily; and new technology is often the answer.

For example, workers in factories, distribution centers, and store warehouses must enter a large amount of data on computers so that the company can keep track of the inventory. Imagine the hard work of keeping lists of thousands of parts or pieces of merchandise on paper! Workers and companies both can benefit from new ways of doing things, because if a company does well, there should be more and better jobs for the workers.

In today's work environment it is important to learn about new equipment, computers, and other technological developments. Technology is developing quickly, and the skills needed to do a job can change quickly as well. Every worker, from the cashier at the corner market to the president of a large company, needs to know about computers and other technological advances in order to compete in today's workplace.

A smart worker will welcome any opportunity to learn something new. Being able to learn new things and being familiar with new technologies makes a worker more valuable, and improves the chances that he or she will be able to take advantage of new opportunities in the future. Training or retraining, either on the job or on your own, is a wise idea for anyone who wants to do well in our ever-changing world.

A. Read the story again and circle each of the vocabulary words listed above.

B. Work with a partner. Discuss the story. Ask if your partner agrees with the ideas in the story. Find out if your partner has any personal experience with the advancement of new technology in the workplace.

Role Play

Imagine that you are training another student to do a procedure you use at work or at home. Choose a procedure and explain to your partner what tools and equipment you use and how you use them. Give the other person instructions for all the steps in the procedure, and teach him or her how to act out the steps. Then have your partner act out the procedure in front of the class. At the same time, explain to the class how to do the procedure.

Coping Skills

Charles and Jose practiced many kinds of communication skills and computer skills when they were talking about their jobs in the warehouse.

Put a ✓ next to the things they did. Give examples.

( )   learned about a new job or procedure

( )   explained steps followed, using the past tense

( )   read the items on a complex computer screen

( )   learned new computer vocabulary

( )   entered information on a shipping computer

( )   selected a command from a menu on the computer

( )   gave instructions to another worker

( )   demonstrated a procedure

( )   helped correct an error by asking about steps in a procedure

With a group, discuss the coping skills above.

What is the most important skill you learned in this unit? Which ones will you use in the future? In what situations will you use them?

Community Assignment

Interview someone in your community who has a job that is not familiar to you. Ask this person all about his or her job: what kind of equipment he or she uses, how to use the equipment, what special information you need to know to do the job, and what kinds of jobs other people have at the same workplace. Ask about how this person learned the job and if there are any special training programs for this job. Then report to your class all the information you learned about this kind of job.

I Found It on the Internet

Have you ever used the Internet? What are some uses of the Internet? What do you think is the most useful or interesting service you can find on the Internet?

Becoming a U. S. Citizen

Samira has applied for United States citizenship. She is anxiously waiting for her interview appointment. She is worried, however, about passing the United States history and government examination for citizenship. She doesn't know what kind of questions to expect, and she does not have a class or any study materials to help her. She is talking to her cousin Genet about it.

Read the story and check true (T) or false (F).

	T	F
1. Samira is a U.S. citizen.	___	___
2. Genet is Samira's sister.	___	___
3. Samira is in a citizenship class.	___	___
4. Samira is worried about her examination.	___	___

Focus on Listening 1

What's the Internet?

Listen to Samira and Genet talking. Then write answers to the following questions and discuss them with the class.

1. Why doesn't Genet have her citizenship booklet?

   _____

2. Why didn't Samira enroll in a citizenship class?

   _____

3. Why can't she go to the library?

   _____

4. What is the Internet?

   _____

5. Does Samira have a computer?

   _____

6. Has Genet ever used the Internet before?

   _____

Look and Listen 👁 👁 👂

A. **Listen to the instructions for an Internet search and look at the computer screen on the next page. Find each of the following items located on the screen and write the name of each item in the correct blank.**

1. toolbar
2. Search button
3. search categories
4. location box
5. name or keywords
6. Go Find It button

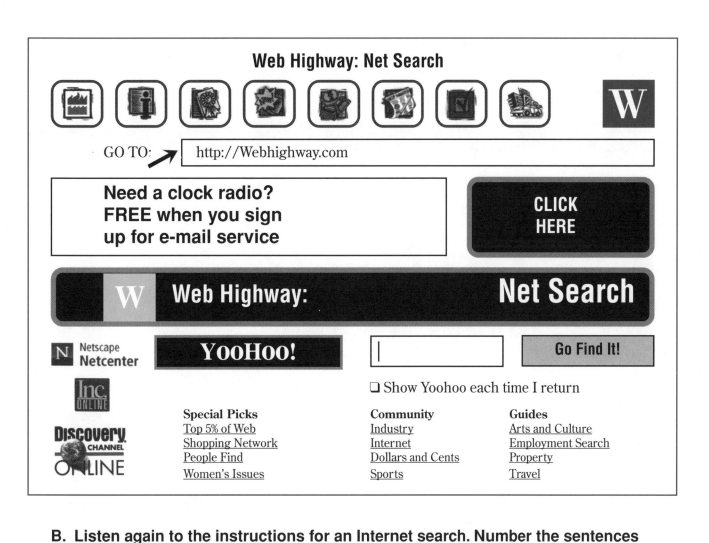

Web Highway: Net Search

GO TO: → http://Webhighway.com

Need a clock radio?
FREE when you sign
up for e-mail service

CLICK HERE

W Web Highway: **Net Search**

N Netscape Netcenter

YooHoo!

Go Find It!

❑ Show Yoohoo each time I return

Inc. ONLINE

Discovery CHANNEL ONLINE

Special Picks
Top 5% of Web
Shopping Network
People Find
Women's Issues

Community
Industry
Internet
Dollars and Cents
Sports

Guides
Arts and Culture
Employment Search
Property
Travel

B. Listen again to the instructions for an Internet search. Number the sentences below in the order that you hear them.

_____ Choose a category from the list in the lower part of the screen.

_____ Type in a name or keywords for what you're looking for in the location box.

_____ Look for the toolbar at the top of your screen.

_____ Click on the **Search** button.

_____ Click on the **Go Find It** button.

_____ If you don't find the category you need, look for the empty box near the top of the screen

on the right side. That's called the location box.

Maybe We Can Find It on the Internet

Words	Meanings
online	using an Internet service
browse	go from place to place and look around
web browser	a service that helps you travel on the Internet
search engine	a service that helps you look for something specific on the Internet
World Wide Web	a part of the Internet where you find most of what you're looking for
web site	a location or destination on the Internet
web page	one screen, or "page," of information on a web site
URL	an Internet address for finding a web site; for example, the URL for the Dominie Press web site is www.dominie.com
www.	an abbreviation for World Wide Web; appears at the beginning of most Internet addresses
.com	an abbreviation at the end of an Internet address; indicates that the web site is a commercial site; that is, it was put on the Internet by a company
password	a word or string of letters that only you know
log on	begin using an Internet service (type in your password and so on)
go to	select an item on the screen; for example, "Go to **Search** on the toolbar."

A. Listen to Genet and Samira as they start their Internet search. Then check true (T) or false (F).

	T	F
1. The first step is to choose a search engine.	_____	_____
2. To use the Internet, you must use an online service.	_____	_____
3. Genet told Samira her password.	_____	_____
4. To go to the web browser, Genet clicked on a button on the screen.	_____	_____
5. To go to the search engine, Genet clicked on a button on the screen.	_____	_____

B. Listen to Genet and Samira's conversation again. Fill in the blanks with the vocabulary words on the previous page. Listen as many times as necessary.

Genet: Okay. Here's what you do. First we have to get _____. I start up my _____

_____, type in my secret _____, and wait.

Samira: What's your secret _____?

Genet: I can't tell you. It's a secret!

Samira: Oh. Then what?

Genet: Then we go to our _____ _____. I just click on this one. We use

"WebHighway," but there are several others.

Samira: Wow. Look at all that stuff. Hey, you get a free clock radio!

Genet: Don't pay attention to that. That's just an _____. They're always trying to sell you

something. See this little button with a picture of an envelope on it?

Samira: Uh-huh.

Genet: That's where we click to send and receive e-mail. But right now we need to choose a

_____ _____.

Samira: What's that?

Genet: It's something that looks up your information on the _____. That's what you need to

_____ for your citizenship information.

Samira: So you need a _____ _____ and a _____ _____?

Genet: Yeah. Okay, we'll use Yoohoo. Now you sit here and type in the _____ for Yoohoo. It's

www.yoohoo.com. Okay. Now we're in business.

Real-Life Reading 1

Using a Search Engine

After Samira and Genet logged on to their search engine, this is the web page they saw.

YooHoo!

What's New   Check E-mail   Personalize   Help

Yoohoo Mail   **Team Yoohoo U.**
free e-mail account   **CLICK HERE**   Join now

options

YooHoo Travel—book a flight, check real-time flight status, browse vacation packages

Shopping - Yellow Pages - White Pages - Maps - Classifieds - Personals - Message Boards - Chat
E-mail - Pager - My Yoohoo - Today's News - Sports - Weather - TV - Stock Quotes - **more...**

Arts & Humanities	**News & Media**	**In the News**
Literature, Photography...	Current Events, Newspapers, TV...	• 30,000 flee Florida fires
		• 4th of July
Business & Economy	**Recreation & Sports**	• World Cup
Companies, Finance, Jobs...	Sports, Travel, Autos, Outdoors...	quarterfinals
		• Wimbledon
Computers & Internet	**Reference**	tennis action
Internet, WWW, Software, Games...	Libraries, Dictionaries, Quotations...	**more...**
Education	**Regional**	
Universities, K-12, College Entrance...	Countries, Regions, US States	**Inside Yoohoo**
Entertainment	**Science**	• Visit our Real Estate Center
Cool Links, Movies, Humor, Music...	Biology, Astronomy, Engineering...	• Yoohoo Avisa—no annual fee
Government	**Social Science**	• Y! Store—build an online store
Military, Politics, Law, Taxes...	Archaeology, Economics, Languages...	in 10 minutes
		• Y! Online—ISP for $14.95
Health	**Society & Culture**	
Medicine, Diseases, Drugs, Fitness...	People, Environment, Religion...	**more...**

Read the choices on the web page. Then find the place to click for each of the items below. Write the number next to the correct place to click on the screen.

1. Click here for help or information about using the service.

2. Click here to receive messages that have been sent to you by electronic mail.

3. Click here to browse a site that contains information about travel and vacations.

4. Click here to read the news about who won the World Cup.

5. Click here to find other news stories that are not on the list given.

6. Click here to find out what people think of a new movie.

7. Click here to find a map of your city.

8. Click here to find a government site with information about applying for U.S. citizenship.

Focus on Information 1

How to Do an Internet Search

A. Read the choices on the next web page. Then read the information below and check true (T) or false (F).

Search [　　　　　　] all of Yoohoo ▼

BUY IT HERE
W Web Highway
N O W

- **Bibliographies** (*3*)
- **Chat** (*2*)
- **Citizenship** (*13*)
- **Conventions and Conferences** (*12*)
- **Countries** (*134*)
- **Documents** (*22*)
- **Embassies and Consulates** (*97*)
- **Ethics** (*5*)
- **Institutes** (*20*)
- **Intelligence** (*41*)
- **International Organizations** (*347*)
- **Law** (*1775*)

- **Military** (*422*)
- **National Symbols and Songs** (*33*)
- **News and Media** (*24*)
- **Politics** (*7078*)
- **Research Labs** (*23*)
- **Statistics** (*37*)
- **Student Government@**
- **Taxes** (*190*)
- **Technology Policy** (*74*)
- **U.S. Government** (*6449*)
- **Indices** (*13*)

After Samira and Genet got online and logged on to their search engine, they chose "U.S. Government" from a list of categories on the Yoohoo home page. Now they will continue to choose the next category and the next one, moving from page to page until they find the specific web site or web page they need. If they go in the wrong direction, they will just click on the Back button on the toolbar, which will take them to the previous page. If they go several pages in the wrong direction, they can click on the Home button, which will take them back to the Yoohoo home page where they started.

On the search screen above, Samira needs to decide which category to choose. If she tries the "Citizenship category," she will find various web sites with the word *citizen* or *civics* in the name of the site. However, she wants to find the official government site for the Immigration and Naturalization Service, or INS, so she decides to click on "U.S. Government."

When Samira clicks on "U.S. Government," she will see a menu that has several choices on it, including "Open this link." She needs to hold down the button on the mouse and move her cursor until the "Open this link" command is highlighted. Then she releases the button on the mouse. She will wait for a few minutes, and the next list of categories will appear.

	T	F
1. A *category* is one choice on a list of different subjects.	___	___
2. A *home page* is the first page of a web site.	___	___
3. The *toolbar* is at the bottom of the screen.	___	___
4. *Citizenship* and *civics* probably have similar meanings.	___	___
5. *Official* and *government* probably have different meanings.	___	___
6. The *Open this link* command will connect you to a new web page.	___	___

B. **Read the choices on the next web page. Then read the information below and check true (T) or false (F).**

Search [＿＿＿＿＿] all of Yoohoo ▼

- **Agencies** (*794*)
- **Budget** (*42*)
- **Documents** (*22*)
- **Embassies and Consulates** (*71*)
- **Employment** (*32*)
- **Executive Branch** (*1626*)
- **Federal Employees** (*54*)
- **Intelligence** (*20*)
- **Judicial Branch** (*202*)

- **Legislative Branch** (*718*)
- **Military** (*2601*)
- **National Security** (*28*)
- **National Symbols and Songs** (*21*)
- **Reengineering** (*4*)
- **Research Labs** (*54*)
- **Statistics** (*27*)
- **U.S. States** (*108*)
- **Indices** (*22*)

When Samira selected "U.S. Government" on the previous web page, another list of categories appeared on the screen. Now she needs to decide the best way to find the Immigration and Naturalization Service web site. She will look at the choices of categories and select the best one for what part of the U.S. government she is looking for. If the vocabulary is familiar, she can make a choice. If she is not sure of the vocabulary, she can try using the location box at the top of the screen and clicking on the Search button.

If Samira chooses a category, it would be best to choose a small one. However, the Agencies category has 794 web sites, and the Executive Branch category has 1,626 web sites! Since it could take a long time to browse through so many sites, she decides to use the location box and the search command instead.

Samira will type *INS* in the location box. There is one more thing she needs to do, though, to limit the number of sites she finds in her search. The box to the right of the Search button says "All of Yoohoo!" She needs to change that command, because if she searches all of Yoohoo, she will find hundreds of sites that have the letters *INS* in their names. In order to search only for sites whose name contains *INS* that are in the U.S. government, she needs to click on the arrow to the right of the command and wait for another choice to come up. When she does this, she will see "Just this category." That's the right choice. She needs to search for *INS* in the category of "U.S. Government" only.

	T	F
1. There is more than one way to get to the INS web site from here.	＿＿	＿＿
2. Samira understands all the vocabulary on the page.	＿＿	＿＿
3. There are categories for all three branches of government.	＿＿	＿＿
4. Samira decided to use the location box and Search button.	＿＿	＿＿
5. She wants to browse all of the possible sites.	＿＿	＿＿
6. She selects "Just this category" to limit the number of matching sites.	＿＿	＿＿

C. Read the matches on the next web page. Then read the information below and check true (T) or false (F).

Yoohoo Category Matches　　　　(1 - 1 of 1)

Government: U.S. Government: Executive Branch: Departments and Agencies:
Department of Justice: Immigration and Naturalization Service (INS)

Buy Software
Up to 40% off!
wired.com

Yoohoo Category Matches　　　　(1 - 7 of 7)

Government: U.S. Government: Executive Branch: Departments and Agencies:
Department of Justice: Immigration and Naturalization Service (**INS**)
- Immigration and Naturalization Service (**INS**) - home page.
- U.S. Immigration and Naturalization Laws and Regulations - official government law site. Contains the Immigration and Nationality Act and 89CFR in their entirety.
- http://www.ins.usdoj.gov/law/

Regional: U.S. States: New York: Government
- **Ins**urance Department - information for consumers and **ins**urance industry. Includes publications, press releases, circular letters, FAQs. Allows licensing requirement packages to be downloaded.
  http://www.**ins**.state.ny.us

Regional U.S. States: Utah: Government
- **Ins**urance Department
  http://www.**ins**-dept.state.ut.us/

Finally! Samira found this group of items after choosing "INS" and "Just this category" on the previous page. The results of the search are given at the top of the screen: 1 matching category and 7 matching sites. But look! Some of these are government sites that have to do with insurance, not with immigration and naturalization. It is still necessary to check several sites to be sure of finding the one you are looking for.

Now you can see all of the categories that the INS belongs in: the Department of Justice and the Executive Branch of the United States government. What a long search it was to find it! Now that Samira has the list of sites on the screen, she can click on any web site name that is underlined, and the search engine will go to that site. The underlined names of web sites that are connected to this page are called *links*. Samira will select the link that says "Immigration and Naturalization Service (INS)—home page."

But wait a minute! There's one more thing that Samira needs to remember. As soon as she clicks on the name of the INS home page and the web site appears on her screen, she needs to write down the Internet address of the site. When the web site appears, the address will be in the location box near the top of her screen. It will say *http://www.ins.usdoj.gov* inside the box. That means that next time all she has to do is start up her web browser and type in the Internet address. The INS home page will appear. Also, if Genet and Jonas want to note the address so that they can use it in the future, they can make a *bookmark* on their computer. Then the Internet address for the INS will be on a list that they can check later, just like when you keep an address book with all your friends' addresses in it.

	T	F
1. All these web sites are related to immigration.	____	____
2. Samira found the link for the INS home page.	____	____
3. The INS web page address is *http://www.ins.usdoj.gov*.	____	____

Real-Life Reading 2

We Found the INS Home Page

A. Read all the items on the INS web page below quickly. Try to remember what you see in the different sections of the page.

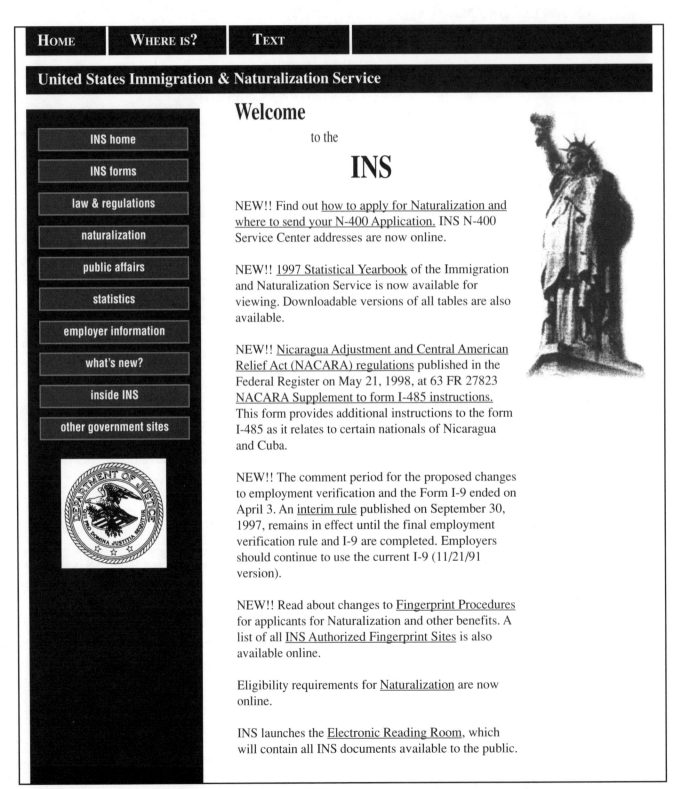

HOME	WHERE IS?	TEXT

United States Immigration & Naturalization Service

- INS home
- INS forms
- law & regulations
- naturalization
- public affairs
- statistics
- employer information
- what's new?
- inside INS
- other government sites

Welcome

to the

INS

NEW!! Find out how to apply for Naturalization and where to send your N-400 Application. INS N-400 Service Center addresses are now online.

NEW!! 1997 Statistical Yearbook of the Immigration and Naturalization Service is now available for viewing. Downloadable versions of all tables are also available.

NEW!! Nicaragua Adjustment and Central American Relief Act (NACARA) regulations published in the Federal Register on May 21, 1998, at 63 FR 27823 NACARA Supplement to form I-485 instructions. This form provides additional instructions to the form I-485 as it relates to certain nationals of Nicaragua and Cuba.

NEW!! The comment period for the proposed changes to employment verification and the Form I-9 ended on April 3. An interim rule published on September 30, 1997, remains in effect until the final employment verification rule and I-9 are completed. Employers should continue to use the current I-9 (11/21/91 version).

NEW!! Read about changes to Fingerprint Procedures for applicants for Naturalization and other benefits. A list of all INS Authorized Fingerprint Sites is also available online.

Eligibility requirements for Naturalization are now online.

INS launches the Electronic Reading Room, which will contain all INS documents available to the public.

1. What is the title of this web page? _____

2. There are two columns of text, or writing, on this web page. One is a list of buttons you can click on to move to different parts of the web site. The other is a list of links available from this page.

   Write "left" or "right" for each of the two columns here.

   Buttons _____   Links _____

3. There are two graphics, or pictures, on this page. Write their names here.

   _____   _____
   (a statue)   (a seal)

4. This is the first page, or home page, of the INS web site. As you move from page to page in this site, you can use the toolbar at the top of this page. Write the three choices on the toolbar here.

   _____   _____   _____

   If your computer could not receive graphics, which menu item would you choose?

   _____

5. *Naturalization* means the application process for becoming a United States citizen. Which two links are of interest to someone who wants to send in an application for naturalization? Copy the names here.

   _____

   _____

6. *Fingerprints* are impressions of a person's fingers used for identification. Which link has information regarding fingerprint procedures for people applying for naturalization? Put a ✓ near the link.

7. *Statistics* means numbers related to immigration. Which new link is available for finding out this kind of information? Mark it with a # symbol.

8. *Eligibility* means rules or requirements about who can apply for naturalization. Which link has this information? Write the name here. _____

9. *Employment verification* means checking a worker's immigration status before giving him or her a job. Which link has information for employers regarding the employment verification rules? Put a $ near the link.

10. Samira has read all of the names of the links, and she hasn't found what she's looking for: naturalization test information. Which button would you try? Read the names of all the buttons on the menu and write the best choice here. _____ Turn the page to see the web page that came up when Samira clicked on this button.

Focus on Reading 1

A. Read the information on this page, and then circle the choice Samira should click on next.

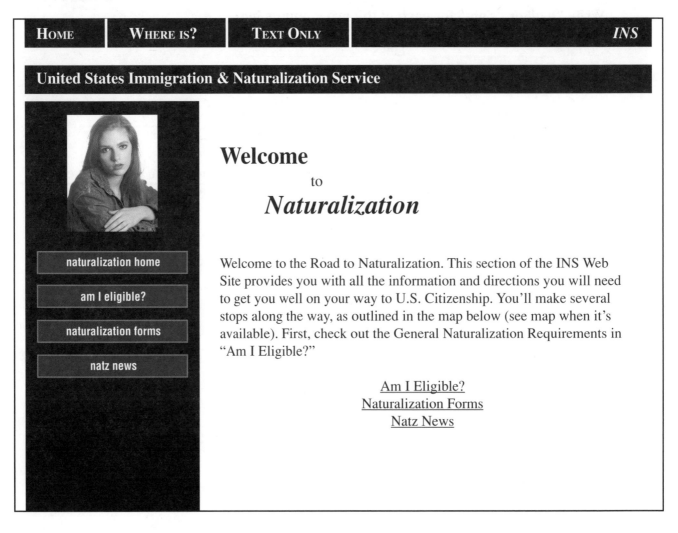

HOME	WHERE IS?	TEXT ONLY		INS

United States Immigration & Naturalization Service

naturalization home

am I eligible?

naturalization forms

natz news

Welcome
to
Naturalization

Welcome to the Road to Naturalization. This section of the INS Web Site provides you with all the information and directions you will need to get you well on your way to U.S. Citizenship. You'll make several stops along the way, as outlined in the map below (see map when it's available). First, check out the General Naturalization Requirements in "Am I Eligible?"

Am I Eligible?
Naturalization Forms
Natz News

B. Study the following words to prepare for reading the next INS web page.

Words	Meanings
download	move information from the Internet to your home computer
software	the electronic language that a computer understands (there are many different kinds of software)
format	the type of software needed to download information from this web site
browse	look for information and read it online (without downloading it onto your home computer)
interactive	a lesson or test on a computer that tells you if your answers are correct

C. Read the information on the next INS web page. Use the vocabulary box to check the meaning of the new words. Then find and circle the words *download, format, browse*, and *interactive*.

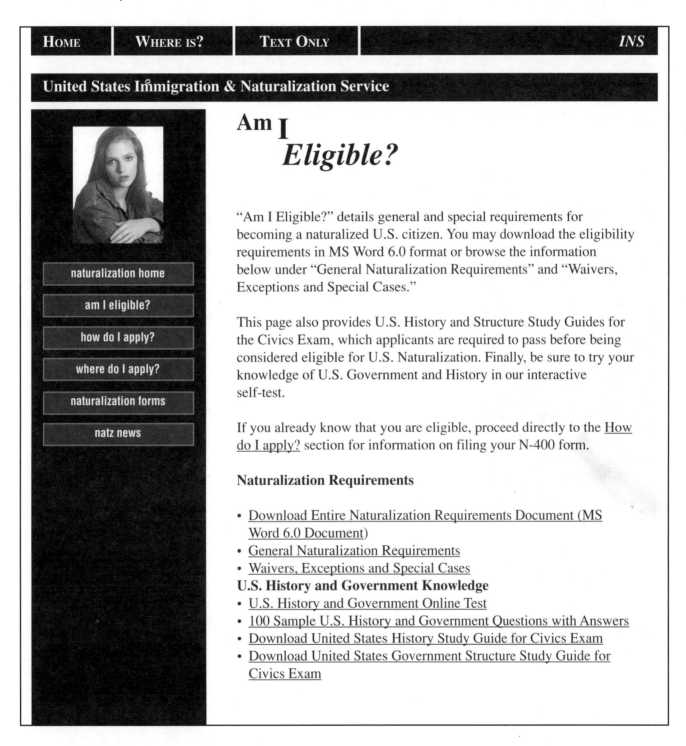

HOME	WHERE IS?	TEXT ONLY	INS

United States Immigration & Naturalization Service

naturalization home

am I eligible?

how do I apply?

where do I apply?

naturalization forms

natz news

Am I *Eligible?*

"Am I Eligible?" details general and special requirements for becoming a naturalized U.S. citizen. You may download the eligibility requirements in MS Word 6.0 format or browse the information below under "General Naturalization Requirements" and "Waivers, Exceptions and Special Cases."

This page also provides U.S. History and Structure Study Guides for the Civics Exam, which applicants are required to pass before being considered eligible for U.S. Naturalization. Finally, be sure to try your knowledge of U.S. Government and History in our interactive self-test.

If you already know that you are eligible, proceed directly to the How do I apply? section for information on filing your N-400 form.

Naturalization Requirements

- Download Entire Naturalization Requirements Document (MS Word 6.0 Document)
- General Naturalization Requirements
- Waivers, Exceptions and Special Cases
U.S. History and Government Knowledge
- U.S. History and Government Online Test
- 100 Sample U.S. History and Government Questions with Answers
- Download United States History Study Guide for Civics Exam
- Download United States Government Structure Study Guide for Civics Exam

D. Put a check mark next to any links on the web page above that would be of interest to Samira. She wants to find out what kind of U.S. history and government questions to expect during her citizenship interview.

Focus on Information 2

U.S. History and Government Online Test

Samira and Genet found this practice test on the INS web site. It's exactly what Samira needs. She can practice answering questions on United States history and government by using any computer with Internet access. When she clicks on the Generate Questions button, she will see five questions. She will answer them by clicking on the correct answer for each one. After she answers five questions, she will see two buttons at the bottom of her screen: *Generate Questions* and *Review Answers*. She can check her answers by clicking on the Review Answers button. Then she can click on Generate Questions again, and five more questions will appear. All of the 100 questions that the INS recommends that she study are included on this practice test.

HOME	WHERE IS?	TEXT ONLY	*INS*

United States Immigration & Naturalization Service

Welcome to the *Naturalization* Self-Test

- naturalization home
- am I eligible?
- how do I apply?
- where do I apply?
- naturalization forms
- natz news

Welcome to the Naturalization Self-Test! To begin the self-test, click the **Generate Questions** button.

This is a test of your knowledge of United States History and the structure of our government. This is not the actual test that you will be given by an INS Officer. It is designed to be used as a study guide only. The study guide test is in multiple choice format. At your interview, the INS Officer will ask you similar questions in an oral format. You will also be evaluated on your ability to speak, read, write, and understand English at your interview.

Generate Questions

Samira has other choices for getting information from the INS web site. She can download a list of 100 sample questions to study or download study guides on U.S. history and U.S. government. To download these materials, she needs to use a computer with the correct software and follow the directions on the screen. There is also one more possibility. She can use the online test, click on the Review Answers button, and then print out one page of questions and answers. All she has to do is click on the Print button on her web browser's toolbar, and the computer will print out one page on paper. Samira is going to ask Jonas if she can use his computer again to answer some practice questions and print out some pages to study. She's very excited about using the Internet to prepare for her INS interview for United States citizenship.

A. Read the information and the web page and answer the following multiple-choice questions.

1. The Naturalization Self-Test has questions on
   a. citizenship requirements
   c. history and government
   b. the N-400 application

2. *Generate Questions* means
   a. answer the questions
   c. ask general questions only
   b. show the questions

3. The questions on this test include
   a. 100 sample questions
   c. 100 years of U.S. history
   b. 100 actual questions

4. At the naturalization interview, the test format will be
   a. oral
   c. multiple choice
   b. written

5. To get a copy of the materials available on the INS web site, Samira can
   a. download questions and study guides
   c. both a and b
   b. print out pages of the online test

6. Samira is planning to
   a. buy a computer
   c. use her cousin's computer
   b. use the computer at the library

B. Go back to pages 134 through 142 and review the selections Genet and Samira made as they clicked through the screens. Make notes of the links they selected and the order in which they selected them. Then compare your notes with a partner.

U.S. Government, Executive Branch, _____

_____

_____

Focus on Reading 2

Participating in Government

When Samira becomes a U.S. citizen, she will be able to vote in all elections. She will also have other ways to use her own *voice* to help make this country better. Many Americans don't realize that their voice *can* be heard. When you have an opinion about something the government is doing, you can contact your representatives in Washington, D.C. and let them know how you feel. If a lot of people do this, it can really bring about changes. All Americans are encouraged to communicate their views to their elected representatives by phone, fax, letter, or e-mail. Make sure your voice is heard!

One way people can easily contact their elected representatives is by the Internet. Most senators and members of Congress have e-mail addresses, and most government offices have web sites. You can even e-mail a letter to the president of the United States by sending your message to: *president@whitehouse.gov.*

If you don't have access to electronic mail at home, you can sometimes send an e-mail by going to the web site of the representative you wish to contact. For example, Samira's congresswoman in the House of Representatives has the following page on her web site. It's very simple to put your name and address in the boxes and then type your letter in the Comments box. After that, click on Record Response, and your message will travel electronically to your representative in Washington, D.C.

Guest Book

Name:

Address:

City:

State:

ZIP:

E-Mail:

Comments:

⬆
⬇

◀ ▶

| Record Reponse | | Clear Entries |

Read the information and the e-mail screen and check true (T) or false (F).

		T	F
1.	It's important for citizens to vote in elections.	____	____
2.	Samira can vote now.	____	____
3.	Elected representatives want to hear the opinions of the people.	____	____
4.	There are many ways to communicate with our representatives in government.	____	____
5.	E-mail is one of the most difficult ways to communicate.	____	____
6.	Americans cannot send their opinions to the president of the United States.	____	____

Crossword Puzzle

Use the new vocabulary from this unit to complete the crossword puzzle.

ACROSS

4 The "information superhighway" that allows for easy access to the Internet
6 A program that connects you to the Internet
7 The blinking line that shows where you are on the computer screen
8 A location or destination on the World Wide Web
9 Using an Internet service
11 The arrow or bar that shows the movement of the mouse on the screen

12 A word or string of letters that only you know

DOWN

1 A line of keys on the computer screen that perform various functions
2 An Internet address
3 A web site that helps you find things on the Internet
5 Look around
7 Push the button on a mouse
10 Begin an Internet session
13 Advertisement

Focus on Communication

Find Someone Who. . .

Circulate around the room and talk to many people. Find someone who fits each of the sentences below and write that person's name in the blank.

_____ has never used a computer.

_____ works with a computer on the job.

_____ has applied for U.S. citizenship.

_____ knows the name of one U.S. senator from our state.

_____ knows how to use a mouse.

_____ heard about the Internet before starting this class.

_____ has tried searching the Internet before.

_____ has sent or received an e-mail message.

Role Play

Imagine that you found some very interesting information on the Internet, and now you are talking to a friend or relative on the telephone. Describe the information you found, how you found it, and what you did with the information. In a group, listen to each person's example and choose the most interesting one. Have that person act out his or her example for the class.

Coping Skills

When Genet and Samira were searching for information on the Internet, they had to try many ways to find what they needed.

A. Put a ✓ next to the things Samira or Genet did. Give examples.

( )   learned new Internet vocabulary

( )   logged on to an Internet browser

( )   ignored the ads on the Internet

( )   conducted an online search

( )   selected a category from a list on the screen

( )   selected a link to search for a web site

( )   narrowed the search by choosing "Just this category"

( )   found the home page they were looking for

( )   found the web pages they were looking for

B. With a group, discuss the coping skills above.

What is the most important skill you learned in this unit? Which ones will you use in the future? In what situations will you use them?

Community Assignment

Go to the library or use any computer that has Internet access. Find out the name of your representative in Congress so that you can contact him or her. Find out the address, telephone number, and fax number for your representative's office. If your representative has an e-mail address or web site, write down the address or location. If you are not able to search for this information online, ask the librarian how you can get the information. Then bring your congressperson's name and contact information to class. In a group, compare the information you found. Discuss what subjects each person in the group would like to write to Congress about.